Buckle Down™

to the
Common Core
State Standards

Mathematics
Grade 8

This book belongs to: _____

ISBN 978-0-7836-7990-7

1CCUS08MM01

13 14 15 16 17 18 19 20

Cover Image: Colorful marbles. © Corbis/Photolibrary

Triumph Learning® 136 Madison Avenue, 7th Floor, New York, NY 10016

Frequently Asked Questions about the Common Core State Standards

What are the Common Core State Standards?

The Common Core State Standards for mathematics and English language arts, grades K–12, are a set of shared goals and expectations for the knowledge and skills that will help students succeed. They allow students to understand what is expected of them and to become progressively more proficient in understanding and using mathematics and English language arts. Teachers will be better equipped to know exactly what they must do to help students learn and to establish individualized benchmarks for them.

Will the Common Core State Standards tell teachers how and what to teach?

No. Because the best understanding of what works in the classroom comes from teachers, these standards will establish *what* students need to learn, but they will not dictate *how* teachers should teach. Instead, schools and teachers will decide how best to help students reach the standards.

What will the Common Core State Standards mean for students?

The standards will provide a clear, consistent understanding of what is expected of student learning across the country. Common standards will not prevent different levels of achievement among students, but they will ensure more consistent exposure to materials and learning experiences through curriculum, instruction, teacher preparation, and other supports for student learning. These standards will help give students the knowledge and skills they need to succeed in college and careers.

Do the Common Core State Standards focus on skills and content knowledge?

Yes. The Common Core State Standards recognize that both content and skills are important. They require rigorous content and application of knowledge through higher-order thinking skills. The English language arts standards require certain critical content for all students, including classic myths and stories from around the world, America's founding documents, foundational American literature, and Shakespeare. The remaining crucial decisions about content are left to state and local determination. In addition to content coverage, the Common Core State Standards require that students systematically acquire knowledge of literature and other disciplines through reading, writing, speaking, and listening.

In mathematics, the Common Core State Standards lay a solid foundation in whole numbers, addition, subtraction, multiplication, division, fractions, and decimals. Together, these elements support a student's ability to learn and apply more demanding math concepts and procedures.

The Common Core State Standards require that students develop a depth of understanding and ability to apply English language arts and mathematics to novel situations, as college students and employees regularly do.

Will common assessments be developed?

It will be up to the states: some states plan to come together voluntarily to develop a common assessment system. A state-led consortium on assessment would be grounded in the following principles: allowing for comparison across students, schools, districts, states and nations; creating economies of scale; providing information and supporting more effective teaching and learning; and preparing students for college and careers.

TABLE OF CONTENTS

Common Core State Standards

TABLE OF CONTENTS

To the Teacher:

Standards Name numbers are listed for each lesson in the table of contents. The numbers in the shaded gray bar that runs across the tops of the pages in the workbook indicate the Standards Name for a given page (see example to the left).

Introduction

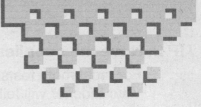

Not a day goes by when you don't use your math skills in some way. You use math to figure out when to wake up in the morning so you can get to school on time, to decide whether you have enough money to buy a new DVD, and to estimate how much room is left on your computer's hard drive. You also use math when you check the temperature on a thermometer to decide whether you need to wear a jacket, and when you compare the amount of time you studied for a test to your score on the test.

This book will help you practice these and many other math skills that you can use in your everyday life, as well as in school. As with anything else, the more you practice these skills, the better you will get at applying them.

Test-Taking Tips

Here are a few tips that will help you on test day.

TIP 1: Take it easy.

Stay relaxed and confident. Because you've practiced these problems, you will be ready to do your best on almost any math test. Take a few slow, deep breaths before you begin the test.

TIP 2: Have the supplies you need.

For most math tests, you will need two sharp pencils and an eraser. Your teacher will tell you whether you need anything else.

TIP 3: Read the questions more than once.

Every question is different. Some questions are more difficult than others. If you need to, read a question more than once. This will help you make a plan for solving the question.

TIP 4: Learn to "plug in" answers to multiple-choice items.

When do you "plug in"? You should "plug in" whenever your answer is different from all of the answer choices or you can't come up with an answer. Plug each answer choice into the problem and find the one that makes sense. (You can also think of this as "working backward.")

TIP 5: Answer open-ended items completely.

When answering short-response and extended-response items, show all your work to receive as many points as possible. Write neatly enough so that your calculations will be easy to follow. Make sure your answer is clearly marked.

TIP 6: Use all the test time.

Work on the test until you are told to stop. If you finish early, go back through the test and double-check your answers. You just might increase your score on the test by finding and fixing any errors you might have made.

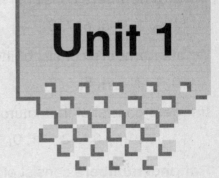

Unit 1

The Number System

Some numbers are convenient to work with, such as 3, −5, or $\frac{1}{3}$. However, some numbers, such as π or $\sqrt{2}$, are not as convenient and can't be represented as easily. You'll experience both types of numbers in your life, so it's important to understand the difference between these types of numbers and learn to estimate them when necessary.

In this unit, you will learn about rational numbers, including how to represent them using a decimal expansion. You will learn about irrational numbers and why they are different from rational numbers. You will also learn to estimate the value of irrational numbers and expressions with irrational numbers. Finally, you will compare and order rational and irrational numbers.

In This Unit

Rational Numbers

Irrational Numbers

Estimating the Value of Expressions

Comparing and Ordering Rational and Irrational Numbers

CCSS: 8.NS.1

Lesson 1: Rational Numbers

The numbers we use every day are part of the **real number system**. The real number system includes the following sets:

Natural numbers consist of the counting numbers.

$\{1, 2, 3, 4, 5, 6, 7, 8, \ldots\}$

Whole numbers consist of the natural numbers and zero.

$\{0, 1, 2, 3, 4, 5, 6, 7, \ldots\}$

Integers consist of the natural numbers, their opposites, and zero.

$\{\ldots, -4, -3, -2, -1, 0, 1, 2, 3, 4, \ldots\}$

Rational numbers consist of the numbers that can be expressed as a fraction, $\frac{a}{b}$, where a and b are both integers and $b \neq 0$. This set includes the integers, decimals that end, and decimals that repeat. Here are a few examples of rational numbers:

$$\frac{2}{1} \qquad \frac{13}{4} \qquad -\frac{25}{100} \qquad \frac{1}{3}$$

Every rational number has a decimal expansion where, in the decimal part, the digits repeat. The decimal can repeat in a single digit, a block of digits, or zeros. To convert a fraction to its decimal expansion, divide the numerator by the denominator until the decimal terminates or repeats.

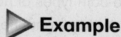 **Example**

Write the decimal expansion of 4.

The whole number 4 can be expressed as a fraction with the denominator 1: $\frac{4}{1}$.

$$\begin{array}{r} 4.0 \\ 1\overline{)4} \\ \underline{-4} \\ 0 \\ \underline{-0} \\ 0 \end{array}$$

Example

Write the decimal expansion of -8.

The integer -8 can be expressed as a fraction with the denominator 1: $-\frac{8}{1}$.

$$\begin{array}{r} -8.0 \\ 1\overline{)-8} \\ \underline{-(-8)} \\ 0 \\ \underline{-0} \\ 0 \end{array}$$

Both 4 and -8 can be extended so that additional zeros are added forever. Therefore, 4 and -8 can also be considered as having repeating decimal expansions. Both 4 and -8 are rational numbers.

CCSS: 8.NS.1

▶ **Example**

Write the decimal expansion of $\frac{9}{16}$.

```
        0.5625
16)9.0000
   − 8
    100
   − 96
      40
    − 32
      80
    − 80
       0
```

$$\frac{9}{16} = 0.5625$$

▶ **Example**

Write the decimal expansion of $\frac{2}{3}$.

```
     0.666
3)2.000
 − 18
   20
 − 18
   20
 − 18
    2
```

Notice the repeating pattern.
$$\frac{2}{3} = 0.\overline{6}$$

The number $\frac{9}{16}$ has a finite decimal expansion that ends. However, you can add zeros to the end of the quotient, 0.56250000…, so it can be considered as having a repeating decimal expansion. The number $\frac{2}{3}$ has a repeating decimal expansion because the number 6 in the quotient repeats forever. Both $\frac{9}{16}$ and $\frac{2}{3}$ are rational numbers.

 TIP: A bar over a number means that the number is repeated endlessly. For example, if you divide $\frac{1}{3}$, the quotient will be 0.3333… with the 3 repeating forever. This number can be represented as: $0.\overline{3}$.

Practice

Directions: For questions 1 through 10, find the decimal expansion for each rational number.

1. $\frac{1}{4}$

.25

_____

2. 88

_____

3. $\frac{2}{9}$

_____

4. −2

_____

5. $\frac{3}{11}$

_____

6. $\frac{5}{1}$

_____

7. $\frac{10}{3}$

_____

8. 1

_____

9. $\frac{5}{12}$

_____

10. $\frac{6}{7}$

_____

CCSS: 8.NS.1

Directions: For questions 11 through 14, find the decimal expansion for each rational number in the given scenario.

11. A supermarket charges $4 for a pack of 3 avocados. What is the decimal expansion for the per-unit price of an avocado at the supermarket?

_____

12. Before 2000, all stocks traded on the New York Stock Exchange were valued using fractions. What was the decimal expansion of a stock listed at $15\frac{31}{32}$?

_____

13. The length of a running track is $\frac{2}{5}$ mile. What is the decimal expansion of this distance?

_____

14. A Pakistani man broke a world record in 2009 by lifting a $160\frac{15}{16}$-pound weight with his right ear. What is the decimal expansion of this weight?

_____

Lesson 2: Irrational Numbers

The real number system includes both rational numbers and irrational numbers. **Irrational numbers** consist of the numbers that cannot be expressed as a fraction of integers. Their decimal expansions go on forever without a repeating pattern. Here are a few examples of irrational numbers:

$$\sqrt{2} \qquad \pi \qquad 45.9492\ldots$$

The decimal expansion of $\sqrt{2}$ is $1.41421356\ldots$. Because the decimal does not end or repeat, it must be irrational.

Even though you cannot find the exact values of irrational numbers, you can approximate their values. To approximate the value of a square root to the closest whole number, find the two perfect squares that the number lies between. The square root of the closer perfect square will be a good approximation.

▷ **Example**

Approximate $\sqrt{10}$ to the nearest whole number.

$\sqrt{10}$ is an irrational number because its decimal value does not end or repeat. However, 10 falls between the perfect squares 9 and 16.

$$\sqrt{9} = 3 \qquad\qquad \sqrt{16} = 4$$

$\sqrt{10}$ is therefore between 3 and 4. Because 10 is closer to 9 than 16, $\sqrt{10}$ is closer to 3 than to 4. The best whole number approximation for $\sqrt{10}$ is 3.

▷ **Example**

Approximate $\sqrt{33}$ to the nearest whole number.

$\sqrt{33}$ is an irrational number because its decimal value does not end or repeat. However, 33 falls between the perfect squares 25 and 36.

$$\sqrt{25} = 5 \qquad\qquad \sqrt{36} = 6$$

$\sqrt{33}$ is therefore between 5 and 6. Because $\sqrt{33}$ is closer to $\sqrt{36}$, its value is closer to 6 than to 5. The best whole number approximation for $\sqrt{33}$ is 6.

 TIP: A perfect square is a number that has an integer as its square root. The first 10 perfect squares are 1, 4, 9, 16, 25, 36, 49, 64, 81, and 100.

CCSS: 8.NS.2

To approximate the value of a square root to the closest tenth without a calculator, find the two perfect squares that the number lies between. Then square the decimal values (to the tenths) between those square roots until you find the closest number.

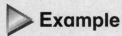 **Example**

Approximate $\sqrt{88}$ to the nearest tenth.

88 falls between the perfect squares 81 and 100.

$\sqrt{81} = 9$ $\qquad\qquad\qquad$ $\sqrt{100} = 10$

Therefore, $\sqrt{88}$ is between 9 and 10. Because 88 is closer to 81 than to 100, $\sqrt{88}$ is closer to 9 than to 10. To find its value to the nearest tenth, square the decimal values to the tenths starting with 9.

$9.0^2 = 81$ $\qquad\qquad\qquad$ $9.3^2 = 86.49$

$9.1^2 = 82.81$ $\qquad\qquad\qquad$ $9.4^2 = 88.36$

$9.2^2 = 84.64$

88 is between 9.3^2 and 9.4^2. 88 is closer to 88.36 than 86.49, so $\sqrt{88}$ is closer to 9.4 than 9.3. $\sqrt{88}$ is 9.4 to the nearest tenth.

You could continue the same process to find the value of an irrational number to the nearest hundredth.

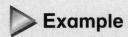 **Example**

Approximate $\sqrt{88}$ to the nearest hundredth.

The value of $\sqrt{88}$ is between 9.3 and 9.4, and it is closer to 9.4. To find the value of $\sqrt{88}$ to the nearest hundredth, square the decimal values to the hundredths starting with 9.39.

$9.39^2 = 88.1721$ $\qquad$ $9.38^2 = 87.9844$ $\qquad$ $9.37^2 = 87.7969$

88 is between 9.38^2 and 9.39^2. Because 88 is closer to 87.9844 than 88.1721, $\sqrt{88}$ is closer to 9.38 than 9.39. The best approximation for $\sqrt{88}$ to the nearest hundredth is 9.38.

Practice

Directions: For questions 1 through 4, find the closest whole number approximation for each irrational number.

1. $\sqrt{5}$

_____

2. $\sqrt{98}$

_____

3. $\sqrt{55}$

_____

4. $\sqrt{77}$

_____

Directions: For questions 5 through 8, find the closest approximation to the nearest tenth for each irrational number.

5. $\sqrt{5}$

_____

6. $\sqrt{98}$

_____

7. $\sqrt{55}$

_____

8. $\sqrt{77}$

_____

9. A square-shaped painting has a length and a width of 1 foot. A line painted diagonally on the painting has a length of $\sqrt{2}$ feet. About how long is the length of the diagonal line, to the nearest tenth of a foot?

_____

CCSS: 8.NS.2

Lesson 3: Estimating the Value of Expressions

To estimate the value of an expression with an irrational number, substitute the irrational number with its approximate decimal value, to the nearest hundredth. Then continue to solve the expression according to the order of operations.

The following table lists 10 common irrational numbers, with their decimal approximations to the hundredths.

Irrational Number	Approximation
$\sqrt{2}$	1.41
$\sqrt{3}$	1.73
$\sqrt{5}$	2.24
$\sqrt{6}$	2.45
$\sqrt{7}$	2.65

Irrational Number	Approximation
e	2.72
$\sqrt{8}$	2.83
π	3.14
$\sqrt{10}$	3.16
$\sqrt{11}$	3.32

▶ **Example**

Approximate $\sqrt{3} + \sqrt{6}$.

The approximate value of $\sqrt{3}$ is 1.73. The approximate value of $\sqrt{6}$ is 2.45.

$1.73 + 2.45 = 4.18$.

Therefore, $\sqrt{3} + \sqrt{6} \approx 4.18$.

▶ **Example**

Approximate π^2.

The approximate value of π is 3.14.

$3.14^2 = 3.14 \times 3.14 = 9.8596$.

Therefore, $\pi^2 \approx 9.86$.

▷ Example

Approximate $e - \sqrt{6} \times \sqrt{3}$.

The approximate value of e is 2.72. The approximate value of $\sqrt{6}$ is 2.45. The approximate value of $\sqrt{3}$ is 1.73.

$2.72 - 2.45 \times 1.73 = 2.72 - 4.2385 = -1.5185$

Therefore, $e - \sqrt{6} \times \sqrt{3} \approx -1.52$.

● Practice

Directions: For questions 1 through 8, find the approximate values of the given expressions, to the nearest hundredth.

1. $\pi - e$

_____

2. $\sqrt{3} - \sqrt{2}$

_____

3. $\sqrt{8} \times \pi$

_____

4. $\sqrt{5} + \sqrt{6} + \sqrt{7}$

_____

5. $10e$

_____

6. $\sqrt{11} - \sqrt{10}$

_____

7. $e^2 + e$

_____

8. $\sqrt{2} + \sqrt{10} \times \pi$

_____

9. The length of the diagonal in quadrilateral A is $5\sqrt{3}$. The length of the diagonal in quadrilateral B is $8\sqrt{2}$. The length of the diagonal in quadrilateral C is $3\sqrt{12}$. What is the approximate sum of the lengths of the diagonals of quadrilaterals A, B, and C?

_____

Lesson 4: Comparing and Ordering Rational and Irrational Numbers

Comparing Rational and Irrational Numbers

To compare rational and irrational numbers, convert them to the same form. Use decimal approximations for irrational numbers. When the numbers are in decimal form, compare the digits from greatest place to least place. Then use $<$, $>$, or $=$ to compare the numbers.

▷ **Example**

Compare $2\frac{4}{5}$ and 2.67.

Convert $2\frac{4}{5}$ to a decimal.

$2\frac{4}{5} = 2.8$

Use $<$, $>$, or $=$ to compare the decimals.
$2.8 > 2.67$

Therefore, $2\frac{4}{5} > 2.67$.

▷ **Example**

Compare 3.1 and π.

Use a decimal approximation for π. $\pi \approx 3.14$.

Use $<$, $>$, or $=$ to compare the decimals.
$3.1 < 3.14$

Therefore, $3.1 < \pi$.

▷ **Example**

Compare π and $\sqrt{10}$.

Use approximations for π and $\sqrt{10}$. $\pi \approx 3.14$ and $\sqrt{10} \approx 3.16$.

Use $<$, $>$, or $=$ to compare the decimals.
$3.14 < 3.16$

Therefore, $\pi < \sqrt{10}$.

19

Ordering Rational and Irrational Numbers

You can use a number line to order rational and irrational numbers. Once all the numbers have been plotted on the number line, they will be in order from least to greatest, from left to right.

 Example

Plot the following numbers on a number line. Then list them in order from least to greatest.

$$\sqrt{6} \qquad \pi \qquad \frac{15}{4} \qquad \sqrt{8} \qquad \sqrt{3}$$

First find decimal approximations for each number.

$$\sqrt{6} \approx 2.45$$

$$\pi \approx 3.14$$

$$\frac{15}{4} = 3.75$$

$$\sqrt{8} \approx 2.83$$

$$\sqrt{3} \approx 1.73$$

Now plot each of these decimals on the number line.

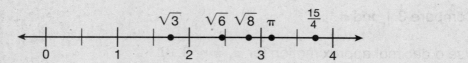

In order from least to greatest, the numbers are $\sqrt{3}$, $\sqrt{6}$, $\sqrt{8}$, π, and $\frac{15}{4}$.

CCSS: 8.NS.2

 Practice

Directions: For questions 1 through 10, compare the two numbers using <, >, or =.

1. $\frac{3}{4}$ and 0.7

_____

2. π and 4

_____

3. $\sqrt{2}$ and 1.44

_____

4. e and $\sqrt{7}$

_____

5. 7.2 and $7\frac{2}{3}$

_____

6. $\sqrt{15}$ and $\sqrt{22}$

_____

7. $\sqrt{8}$ and π

_____

8. $\frac{8}{3}$ and e

_____

9. $\frac{2}{9}$ and $\frac{3}{11}$

_____

10. 3 and $\sqrt{9}$

_____

Directions: For questions 11 through 15, use the number line to order the rational and irrational numbers from least to greatest.

11. $\sqrt{2}, -0.9, \frac{1}{8}, \sqrt{3}, -\sqrt{2}$

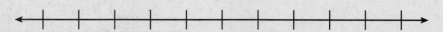

12. $\sqrt{8}, \pi, e, \sqrt{15}, \sqrt{18}$

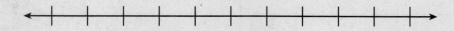

13. $\frac{5}{2}, \sqrt{7}, \pi, 1.\overline{1}, \sqrt{1}$

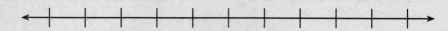

14. $e, \frac{2}{8}, \sqrt{5}, \frac{0}{99}, \sqrt{2}$

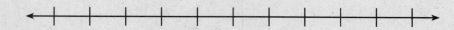

15. $-\frac{10}{9}, -e, -2.09, -0.5^2, -\sqrt{3}$

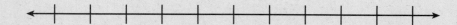

16. A carpenter uses four wooden boards with different lengths. The lengths are 4 ft, $2\sqrt{3}$ ft, $\pi + \sqrt{2}$ ft, and 1.9^2 ft. What are the lengths of the boards in order from shortest to longest?

_____

Unit 1 Practice Test

Read each question. Choose the correct answer.

1. Which whole number approximation is closest to the value of $\sqrt{140}$?

   A. 10

   B. 11

   C. 12

   D. 14

2. What is the decimal expansion for $\frac{5}{6}$?

   A. 0.56

   B. 0.8333333

   C. $0.8\overline{3}$

   D. $0.\overline{83}$

3. Which is the most accurate approximation for $\sqrt{8} - \sqrt{2}$?

   A. 1.42

   B. 2

   C. 2.45

   D. 4.24

4. What is the decimal expansion for $\frac{7}{12}$?

   A. 0.583

   B. $0.58\overline{3}$

   C. $0.5\overline{83}$

   D. $0.\overline{583}$

5. Which is the most accurate approximation for $\pi + e$?

   A. 0.42

   B. 5

   C. 5.86

   D. 9

6. Which statement correctly compares $\sqrt{11}$ and 3.3?

   A. $\sqrt{11} < 3.3$

   B. $3.3 = \sqrt{11}$

   C. $3.3 > \sqrt{11}$

   D. $\sqrt{11} > 3.3$

7. What whole number approximation is closest to the value of $\sqrt{31}$?

_____

8. The length of string A is $\sqrt{18}$ in. The length of string B is 5.42 in. Compare the lengths of the two strings using the symbol $<$, $>$, or $=$.

_____

9. Approximate the sum of $e + \sqrt{6}$.

_____

10. What is the decimal expansion for $\frac{8}{9}$?

_____

11. Approximate the product of $\sqrt{3} \times 0.\overline{3}$.

_____

12. Adam's height is $\sqrt{2}$ yd. His father's height is 1.91 yd. Compare the heights of Adam and his father using $<$, $>$, or $=$.

_____

13. What is the decimal expansion for $\frac{5}{11}$?

_____

14. What whole number is closest to the value of 10π?

_____

15. Mindy wrote the following four numbers in her notebook.

$$7.31038... \quad \sqrt{38} \quad \frac{51}{10} \quad \sqrt{31}$$

What are Mindy's four numbers in order from least to greatest? You may use the following number line to compare them.

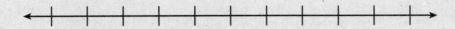

_____

16. A scientist calculates that the distance of a star is $\sqrt{97}$ light-years away from Earth. The scientist wants to approximate this distance to the nearest hundredth of a light-year. What number should the scientist use for her approximation, in light-years?

_____

17. Alfonso needs to approximate the value of the following expression.

$$\pi + \sqrt{50} \times \sqrt{15}$$

What value could Alfonso use for his approximation?

_____

18. Geoffrey wrote the following four numbers on his classroom's blackboard.

$$\sqrt{17} \quad \pi \quad \sqrt{5} \quad e$$

What are Geoffrey's four numbers in order from greatest to least? You may use the following number line to compare them.

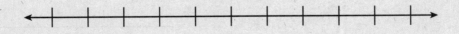

_____

19. What whole number is closest to the value of $\sqrt{85}$?

_____

20. Approximate the value of e^2.

_____

21. What is the decimal expansion for $\frac{27}{3}$?

_____

22. What is the decimal expansion for $\frac{8}{66}$?

_____

23. Approximate the quotient of $\frac{\sqrt{99}}{\sqrt{15}}$.

_____

24. Clay needs to round e to the nearest hundredth. What value should Clay determine for his approximation?

_____

25. Tameka needs to approximate the value of the following expression.

$$5\sqrt{2} - 3\sqrt{2} \times 2.\overline{9}$$

What approximation can Tameka use for the value of the expression?

_____

26. What is the decimal expansion for $\frac{8}{5}$?

_____

27. Nadine needs to approximate the value of $\sqrt{96}$ to the nearest tenth. What number should she use for her approximation?

_____

28. Leslie needs to approximate the value of the following expression.

$$\sqrt{19} + \sqrt{20} + \sqrt{21}$$

What value could Leslie use for his approximation?

_____

29. William plots an irrational number on the following number line.

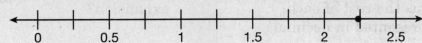

If William's irrational number is a square root value, what number could he have plotted?

_____

30. Write the following three irrational numbers in order from least to greatest. You can use the number line to help put them in order.

$$\sqrt{6} \qquad 3\sqrt{2} \qquad 2\sqrt{3}$$

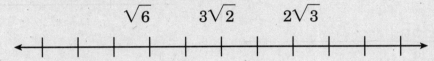

_____

31. The lengths of two snails are shown in the following picture.

4.3203... in. $\sqrt{24}$ in.

Compare the lengths of the snails using the symbol <, >, or =.

_____

32. The cost of 3 bananas at the market is \$1. What is the cost of each banana, represented in decimal expansion?

_____

33. Between which two whole numbers is $\sqrt{110}$?

_____

34. Approximate the difference of the following expression.

$$\sqrt{10} - \pi$$

_____

35. The weight of a honeydew melon is $\sqrt{29}$ pounds. The weight of a pumpkin is 2π pounds. Compare the weights of the melon and the pumpkin using the symbols $<$, $>$, or $=$.

_____

36. What is the value of $\sqrt{3}$ to the nearest hundredth?

_____

37. What is the value of $\frac{17}{15}$ in decimal expansion?

_____

38. Between which two whole numbers is $\sqrt{35}$?

_____

39. Leroy writes a fraction with a numerator of 1. If his denominator is a whole number less than 6, and the decimal expansion of his fraction repeats with a digit other than 0, what is his denominator?

_____

40. Aubrey needs to approximate the value of the following expression.

$$\sqrt{145} - \sqrt{15} \times \pi$$

What approximation can Aubrey use for the value of the expression?

_____

41. What is the value of $\frac{1}{37}$ in decimal expansion?

_____

42. Approximate the product.

$$e \times \sqrt{10}$$

_____

43. What is the value of $\frac{1}{18}$ in decimal expansion?

_____

44. What whole number is closest to the value of $\sqrt{117}$?

_____

45. Oliver needs to compare $\sqrt{2}$ and $\frac{\pi}{2}$. Compare Oliver's numbers using the symbols $<$, $>$, or $=$.

_____

46. The circumference of a circle is determined by multiplying its diameter by π. If the diameter of a circle is 5 ft, what is an approximation for its circumference?

_____

47. What is the value of $\sqrt{108}$ to the nearest hundredth?

_____

48. What is the decimal expansion of $\frac{15}{24}$?

_____

49. Jin needs to approximate the value of the following expression.

$$\sqrt{50} + \sqrt{75} + \sqrt{120}$$

What value could Jin use for his approximation?

_____

50. What is the value of $\sqrt{40}$ to the nearest tenth?

_____

51. What is the decimal expansion of $\frac{3}{1000}$?

_____

52. Deanna is trying to understand whether a given number is rational or irrational.

    Part A
    Is $\sqrt{2}$ a rational or irrational number?

    _____

    Part B
    Explain how you know whether $\sqrt{2}$ is rational or irrational.

    _____

    _____

    _____

53. Xavier needs to find an approximation for the value of $\sqrt{20}$ to make a diagonal truss for a wall.

    Part A
    What is the value of $\sqrt{20}$ to the nearest tenth?

    _____

    Part B
    Explain how you found your answer without using a calculator.

    _____

    _____

    _____

Unit 2

Expressions and Equations

Simplifying expressions and solving equations are important tools that we use to describe the world around us. Expressions can be used to represent a quantity within an equation. An equation relates two expressions, and can model the relationship between changing quantities. A buyer for a department store uses equations to relate how much money is available to spend and the quantities of merchandise the store can buy. A farmer creates equations to model the changing height of his crops over time.

In this unit, you will simplify expressions with exponents and numbers in scientific notation. You will solve equations by writing equivalent equations. You will find the solutions for equations with two variables and learn how to graph them. You will see how linear equations can represent proportional relationships. Finally, you will learn about the different ways to solve systems of equations.

Lesson 5: Exponents

An exponent (or power) shows how many times a base number appears as a factor.
To evaluate 5^4, multiply four fives together.

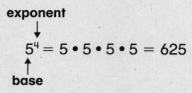

$$5^4 = 5 \bullet 5 \bullet 5 \bullet 5 = 625$$

Evaluating Exponents

Here are a few reminders for evaluating exponents.

- A base with an exponent of 0 equals 1.

$$10^0 = 1 \qquad 5^0 = 1 \qquad 25{,}000{,}000^0 = 1$$

- A base with an exponent of 1 equals the base number.

$$10^1 = 10 \qquad 5^1 = 5 \qquad 25{,}000{,}000^1 = 25{,}000{,}000$$

- A positive base with a positive exponent equals a positive number.

$$4^2 = 16 \qquad 5^3 = 125 \qquad \left(\frac{1}{2}\right)^2 = \frac{1}{4}$$

- A negative base with an even exponent equals a positive number.

$$(-3)^2 = (-3) \bullet (-3) = 9$$

- A negative base with an odd exponent equals a negative number.

$$(-3)^3 = (-3) \bullet (-3) \bullet (-3) = -27$$

- A base with a negative sign in front equals a negative number.

$$-3^3 = -(3 \bullet 3 \bullet 3) = -27 \qquad -9^2 = -(9 \bullet 9) = -81$$

- A base with a negative exponent equals the reciprocal of the base with a positive exponent.

$$5^{-3} = \frac{1}{5^3} = \frac{1}{125} \qquad (-8)^{-3} = \frac{1}{(-8)^3} = -\frac{1}{512}$$

CCSS: 8.EE.1

Multiplying Numbers with Exponents

To multiply numbers with exponents that have the same base, add the exponents and keep the base the same.

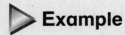 **Example**

Simplify: $2^3 \cdot 2^7$

$$2^3 \cdot 2^7 = (2 \cdot 2 \cdot 2) \cdot (2 \cdot 2 \cdot 2 \cdot 2 \cdot 2 \cdot 2 \cdot 2)$$

$$= 2^{3+7}$$

$$= 2^{10}$$

The simplified version of $2^3 \cdot 2^7$ is 2^{10}.

Dividing Numbers with Exponents

To divide numbers with exponents that have the same base, subtract the exponent of the denominator from the exponent of the numerator and keep the base the same.

Example

Simplify: $\dfrac{3^8}{3^5}$

$$\frac{3^8}{3^5} = \frac{3 \cdot 3 \cdot 3 \cdot 3 \cdot 3 \cdot 3 \cdot 3 \cdot 3}{3 \cdot 3 \cdot 3 \cdot 3 \cdot 3}$$

$$= 3^{8-5}$$

$$= 3^3$$

The simplified version of $\dfrac{3^8}{3^5}$ is 3^3.

Practice

Directions: For questions 1 through 8, evaluate the exponents.

1. $4^2 =$ _____

2. $-8^3 =$ _____

3. $7^{-3} =$ _____

4. $0^1 =$ _____

5. $12^1 =$ _____

6. $-6^4 =$ _____

7. $(-22)^0 =$ _____

8. $(-3)^{-3} =$ _____

Directions: For questions 9 through 22, simplify the multiplication or division expressions.

9. $3^2 \times 3^4 =$ _____

10. $6^2 \times 6 =$ _____

11. $7^5 \times 7^{-2} =$ _____

12. $6^0 \times 6^0 =$ _____

13. $5^3 \times 5^{-6} =$ _____

14. $8^3 \times 8^2 \times 8^{-7} =$ _____

15. $4^{-4} \times 4^2 \times 4^6 =$ _____

16. $\dfrac{3^9}{3^6} =$ _____

17. $\dfrac{8^2}{8^{-2}} =$ _____

18. $\dfrac{8^1}{8^3} =$ _____

19. $\dfrac{1^{10}}{1^8} =$ _____

20. $\dfrac{5^{-2}}{5^2} =$ _____

21. $\dfrac{8^7}{8^3} =$ _____

22. $\dfrac{9^{14}}{9^{12}} =$ _____

Lesson 6: Scientific Notation

Scientific notation is used to represent very large or very small numbers. The following table shows the first five positive and negative powers of 10.

Powers of 10

Positive	Negative
$10^1 = 10$	$10^{-1} = 0.1$
$10^2 = 100$	$10^{-2} = 0.01$
$10^3 = 1,000$	$10^{-3} = 0.001$
$10^4 = 10,000$	$10^{-4} = 0.0001$
$10^5 = 100,000$	$10^{-5} = 0.00001$
and so on…	and so on…

Changing from Standard Form to Scientific Notation

A number is written in scientific notation as the product of a number greater than or equal to 1 but less than 10, and a power of 10. Follow these steps to change a number from standard form to scientific notation.

Step 1: Move the decimal point to the left or right until you have a number greater than or equal to 1 but less than 10.

Step 2: Count the number of places you moved the decimal point to the left or right and use that number as the positive or negative power of 10.

Step 3: Multiply the decimal (from Step 1) by the power of 10 (from Step 2).

▷ **Example**

Write 5,167,800 in scientific notation.
Move the decimal point 6 places to the left.
5.167800.

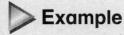

Since the decimal point moved 6 places to the left, multiply by 10^6.
$5.1678 \cdot 10^6$
Therefore, $5,167,800 = 5.1678 \cdot 10^6$.

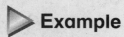 **Example**

Write 0.00000364 in scientific notation.
Move the decimal point 6 places to the right.
0.000003.64

Since the decimal point moved 6 places to the right, multiply by 10^{-6}.
$3.64 \cdot 10^{-6}$
Therefore, $0.00000364 = 3.64 \cdot 10^{-6}$.

Changing from Scientific Notation to Standard Form

To change a number written in scientific notation with a positive power of 10 to standard form, move the decimal point to the right. The power of ten shows the number of places the decimal point will move.

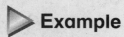 **Example**

Write $4.91 \cdot 10^{8}$ in standard form.
Since the exponent is positive, move the decimal point 8 places to the right and add zeros.
$4.91 \cdot 10^{8} = 4.91000000. = 491,000,000$

Therefore, $4.91 \cdot 10^{8} = 491,000,000$.

To change a number written in scientific notation with a negative power of 10 to standard form, move the decimal point to the left. The power of ten shows the number of places the decimal point will be moved.

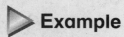 **Example**

Write $3.135 \cdot 10^{-4}$ in standard form.
Since the exponent is negative, move the decimal point 4 places to the left and add zeros.
$3.135 \cdot 10^{-4} = 0.0003.135 = 0.0003135$

Therefore, $3.135 \cdot 10^{-4} = 0.0003135$.

 TIP: Most calculators use 'E' to represent $\cdot 10^{n}$. For instance, the number 4,150,000 can be written in scientific notation as $4.15 \cdot 10^{6}$, and would appear in a calculator as 4.15E6.

CCSS: 8.EE.3, 8.EE.4

Multiplying Numbers in Scientific Notation

When multiplying two numbers in scientific notation, multiply their coefficients and add their exponents. You may need to convert the product into scientific notation if the coefficient is smaller than 1, or 10 or greater.

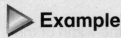 **Example**

Multiply: $(5.1 \cdot 10^4) \times (2.3 \cdot 10^6)$

Multiply the coefficients: $5.1 \times 2.3 = 11.73$

Add the powers of 10: $4 + 6 = 10$

Check to be sure the product is in scientific notation.

$$(5.1 \cdot 10^4) \times (2.3 \cdot 10^6) = 11.73 \cdot 10^{10}$$
$$= 1.173 \cdot 10^{11}$$

Therefore, $(5.1 \cdot 10^4) \times (2.3 \cdot 10^6) = 1.173 \cdot 10^{11}$

Dividing Numbers in Scientific Notation

When dividing two numbers in scientific notation, divide their coefficients and subtract their exponents. You may need to convert the quotient into scientific notation if the coefficient is smaller than 1, or greater than 10.

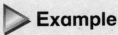 **Example**

Divide: $\frac{(3.9 \cdot 10^8)}{(6.5 \cdot 10^{-4})}$

Divide the coefficients: $\frac{3.9}{6.5} = 0.6$

Subtract the powers of 10: $8 - (-4) = 12$

Check to be sure the quotient is in scientific notation.

$$\frac{(3.9 \cdot 10^8)}{(6.5 \cdot 10^{-4})} = 0.6 \cdot 10^{12}$$
$$= 6.0 \cdot 10^{11}$$

Therefore, $\frac{(3.9 \cdot 10^8)}{(6.5 \cdot 10^{-4})} = 6.0 \cdot 10^{11}$

Using Scientific Notation to Estimate

You can use scientific notation to estimate certain measurements and how much bigger one measurement is in terms of another.

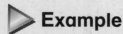 **Example**

How many times larger is the circumference of the Earth than the width of the United States?

First you need to estimate the width of the United States, which is about $3.0 \cdot 10^3$ miles. The circumference of the Earth is about $3.0 \cdot 10^5$ miles. To find how many times larger, divide the circumference of the Earth by the width of the United States.

$$\frac{3.0 \cdot 10^5}{3.0 \cdot 10^3} = 1 \cdot 10^{5-3} = 1 \cdot 10^2 = 100$$

The circumference of the Earth is about 100 times the width of the United States.

CCSS: 8.EE.3, 8.EE.4

⬤ Practice

Directions: For questions 1 through 6, write each number in scientific notation.

1. 0.005

   _____

2. 61,713

   _____

3. 874

   _____

4. 0.00063

   _____

5. 23,009,040

   _____

6. 0.99714

   _____

Directions: For questions 7 through 12, write each number in standard form.

7. $3.06 \cdot 10^4$

   _____

8. $4.00902 \cdot 10^7$

   _____

9. $1.87 \cdot 10^{-3}$

   _____

10. $9.634 \cdot 10^3$

    _____

11. $7.208 \cdot 10^{-9}$

    _____

12. $6.2 \cdot 10^{-5}$

    _____

Directions: For questions 13 through 18, compute. Leave your answer in scientific notation.

13. $\dfrac{(4.8 \bullet 10^2)}{(2.4 \bullet 10^6)} =$ _____

14. $(6.2 \bullet 10^{-3}) \bullet (-3.17 \bullet 10^7) =$ _____

15. $\dfrac{(1.24 \bullet 10^9)}{(3.1 \bullet 10^7)} =$ _____

16. $(2.476 \bullet 10^4) \bullet (6.58 \bullet 10^5) =$ _____

17. $\dfrac{(5.6 \bullet 10^{-4})}{(2.24 \bullet 10^{-5})} =$ _____

18. $(9.18 \bullet 10^{11}) \bullet (2.2 \bullet 10^{-1}) =$ _____

Directions: For questions 19 through 21, solve the problems by computing with scientific notation.

19. The speed of light is about $3 \bullet 10^8$ meters per second. About how far can light travel in 2 hours? (Hint: Convert time to seconds.)

    _____

20. There are 39.37 inches in 1 meter. How many inches are in $8 \bullet 10^4$ meters?

    _____

21. A liter of water has about $3.3 \bullet 10^{22}$ molecules in it. About how many molecules would there be in 3,500 liters of water?

    _____

CCSS: 8.EE.7a, 8.EE.7b

Lesson 7: Linear Equations in One Variable

A variable is a symbol, such as *s* or *x*, that represents an unknown number. You can solve linear equations in one variable using equivalent equations.

Equivalent equations are equations that have the same solution. Two equations have the same solution if you can substitute the same number for their variable, and arrive at a true statement. You can create equivalent equations by adding or subtracting the same term from both sides of an equation. You can also create equivalent equations by multiplying or dividing every term in an equation by the same term.

 Example

Solve the following equation for *x*.

$$-12x = 132$$

Since division is the inverse operation of multiplication, divide both sides of the equation by -12.

$$\frac{-12x}{-12} = \frac{132}{-12}$$

$$x = -11$$

Check your answer by substituting -11 for *x* in the original equation.

$$-12x = 132$$

$$-12 \times (-11) \stackrel{?}{=} 132$$

$$132 = 132 \checkmark$$

Since the substitution makes the equation true, the solution is $x = -11$.

In this example an equivalent equation was made by dividing both sides of the original equation by -12. Since you are performing the same operation on both sides of the equation, the solution of the new equation is the same as the solution of the original equation.

Some equations may have no solution or an infinite number of solutions.

▷ Example

Solve the following equation for *x*.

$$12 + 3x = 3x - 3$$

Since subtraction is the inverse of addition, subtract 12 from both sides of the equation.

$$12 + 3x - 12 = 3x - 3 - 12$$
$$3x = 3x - 15$$

Then subtract 3*x* from both sides of the equation, to eliminate the variable from one side of the equation.

$$3x - 3x = 3x - 3x - 15$$
$$0 = -15$$

Because $0 \neq -15$, there is no value for *x* that would make the equation true. Therefore, there is no solution.

▷ Example

Solve the following equation for *x*.

$$5x + 4 - x = 2 + 4x + 2$$

First, combine the like terms.

$$4x + 4 = 4 + 4x$$

Then subtract 4*x* from both sides of the equation, to eliminate the variable from one side of the equation.

$$4x - 4x + 4 = 4 + 4x - 4x$$
$$4 = 4$$

Because $4 = 4$, every value for *x* will make the equation true. Therefore, there are an infinite number of solutions for *x*.

CCSS: 8.EE.7a, 8.EE.7b

The **distributive property** relates multiplication to addition or subtraction. To find the product of a number and a sum or difference, multiply the outside number by both terms in the sum or difference.

$a(b + c) = a \cdot b + a \cdot c$

$10(4 + 1) = 10 \cdot 4 + 10 \cdot 1$

$10(5) = 40 + 10$

$50 = 50$

$a(b - c) = a \cdot b - a \cdot c$

$\frac{3}{4}(20 - 12) = \frac{3}{4} \cdot 20 - \frac{3}{4} \cdot 12$

$\frac{3}{4}(8) = 15 - 9$

$6 = 6$

▶ ## Example

Simplify the following equation using the distributive property, and then solve for *x*.

$$3(4x - 5) + 2 = -4\left(\frac{3}{2} - 4x\right)$$

Multiply each term in the parentheses by the term outside the parentheses. Then combine like terms.

$$3 \cdot 4x - 3 \cdot 5 + 2 = -4 \cdot \frac{3}{2} - (-4) \cdot 4x$$

$$12x - 15 + 2 = -6 + 16x$$

$$12x - 13 = -6 + 16x$$

$12x - 12x - 13 = -6 + 16x - 12x$ Add $-12x$ to each side of the equation.

$$-13 = -6 + 4x$$

$-13 + 6 = -6 + 4x + 6$ Add 6 to each side of the equation.

$$-7 = 4x$$

$-\frac{7}{4} = \frac{4x}{4}$ Divide both sides of the equation by 4.

$$-\frac{7}{4} = x$$

43

▷ **Example**

Simplify the following equation using the distributive property, and then solve for *x*.

$$-2(-5x - 12) = 5(2x + 4) + 4$$

Multiply each term in the parentheses by the term outside the parentheses. Then combine like terms.

$$-2 \cdot (-5x) - (-2) \cdot (12) = 5 \cdot 2x + 5 \cdot 4 + 4$$
$$10x - (-24) = 10x + 20 + 4$$
$$10x + 24 = 10x + 24$$
$$10x - 10x + 24 = 10x - 10x + 24 \qquad \textbf{Subtract 10}\textbf{\textit{x}}\textbf{ from both sides}$$
$$\textbf{of the equation.}$$
$$24 = 24$$

Because 24 = 24, every value for *x* will make the equation true. Therefore, there are an infinite number of solutions for *x*.

▷ **Example**

Simplify the following equation using the distributive property, and then solve for *x*.

$$3(-2x + 4) = -6x - 11$$

Multiply each term in the parentheses by the term outside the parentheses. Then combine like terms.

$$3(-2x + 4) = -6x - 11$$
$$3 \cdot (-2x) + 3 \cdot 4 = -6x - 11$$
$$-6x + 12 = -6x - 11$$
$$-6x + 12 + 6x = -6x - 11 + 6x \qquad \textbf{Add 6}\textbf{\textit{x}}\textbf{ to both sides of the equation.}$$
$$12 = -11$$

Because 12 ≠ −11, there is no value for *x* that would make the equation true. Therefore, there is no solution.

 TIP: You can check your solution to an equation by substituting the solution for the variable in the original equation. If your solution is "infinite solutions," you can try several different numbers for the variable. If your solution is correct, they will each make the equation true. If your solution is "no solution," no value you substitute for the variable will make the equation true.

CCSS: 8.EE.7a, 8.EE.7b

⬤ Practice

Directions: For questions 1 through 14, solve for the variable in each equation. If there is no solution, write "No solution." If there are an infinite number of solutions, write "Any value makes the equation true."

1. $\frac{z}{4} = -7$

   _____

2. $5b - 13 = 37$

   _____

3. $3s = -21$

   _____

4. $12 - 12n = -48$

   A. $n = 5$
   B. $n = 3$
   C. $n = -3$
   D. $n = -5$

5. $-\frac{1}{2}x + 4 = -3$

   _____

6. $8x - 3 = 4(2x - 3)$

   _____

7. $5(4x - 2) = -10(-2x + 1)$

   _____

8. $k - 18 = -32$

   A. $k = -50$
   B. $k = -14$
   C. $k = 14$
   D. $k = 50$

CCSS: 8.EE.7a, 8.EE.7b

9. $3\left(\frac{2}{3}z - 1\right) = 5$

_____

12. $\frac{2}{3}j - \frac{1}{6} = \frac{1}{2}$

_____

10. $12(0.5x - 5) = -6(-x + 10)$

_____

13. $3 + 3m + 1 = 2m + 4 + m$

_____

11. $a + 6 = 18$

_____

14. $4(3y - 1) = 6(3 - y)$

_____

15. While solving an equation for x, Lewis is left with the equation $2 = 4$. Explain why Lewis can be certain that the equation does not have a solution.

_____

_____

_____

CCSS: 8.EE.7a, 8.EE.7b

Lesson 8: Using One-Variable Linear Equations to Solve Problems

You can often solve word problems by writing an equation that represents the scenario. To do this, you need to translate the problem into an algebraic equation. Then, you can solve for the unknown variable to find the answer to the problem.

 Example

Traci bought boxes of ceramic tiles to tile her kitchen floor. She bought a total of 210 ceramic tiles. Each box has 15 ceramic tiles. How many boxes of ceramic tiles did Traci buy?

You can write the following equation to represent this situation, where b is the number of boxes of ceramic tiles Traci bought.

$$15b = 210$$

Solve the equation for b.

$$\frac{15b}{15} = \frac{210}{15}$$ Divide both sides of the equation by 15.

$$b = 14$$

Traci bought 14 boxes of ceramic tiles.

 Example

One bag of trail mix has 3 ounces of raisins and some peanuts. Rob buys 6 bags of trail mix and has 48 ounces of trail mix altogether. How many ounces of peanuts are in each bag of trail mix?

You can write the following equation to represent this situation, where p is the ounces of peanuts in each bag of trail mix.

$$6(3 + p) = 48$$

Solve the equation for p.

$$6(3 + p) = 48$$ **Use the distributive property.**
$$6 \cdot 3 + 6 \cdot p = 48$$
$$18 + 6p = 48$$
$$6p = 30$$ **Divide both sides of the equation by 6.**
$$p = 5$$

There are 5 ounces of peanuts in each bag of trail mix.

Practice

Directions: For questions 1 through 5, solve the word problem by writing an equation, and solving it.

1. At Antonio's Pizza, a pepperoni pizza costs $6.95. Extra toppings are available for $0.50 each. If Greg bought a pizza for $8.45, how many extra toppings, *t*, did he order?

   A.   1.5

   B.   3

   C.   5

   D. 150

2. Nikki bought a bag of jelly beans. She divided the jelly beans equally among herself and three friends. Each person received 24 jelly beans. How many jelly beans, *j*, were in the bag?

   A.   6

   B.   8

   C. 72

   D. 96

3. Liz spent a total of $44.88 at the mall. She has $7.62 left. How much money, *m*, did Liz have when she arrived at the mall?

   _____

4. Melissa has 6 times as many quarters as Michelle. Together, they have a total of 896 quarters. How many quarters, *q*, does Michelle have?

   _____

5. Jim's family went on vacation and rented a car. The rental car agency charged $64.75 plus an additional $0.03 for each mile the car was driven. If Jim's family paid a total of $71.14 for the car rental, how many miles did the family drive the car?

   _____

   Explain how you set up an equation to solve this word problem.

   _____

   _____

   _____

CCSS: 8.EE.2

Lesson 9: Solving Non-Linear One-Variable Equations

Roots of Numbers

The nth root of a number x is represented by the symbol $\sqrt[n]{x}$. The symbol $\sqrt{}$ is called the radical sign, n is called the index (the root to take), and x is called the radicand. Finding the nth root of a number is the inverse of raising a number to the nth power. This means that, $\sqrt[n]{x} = r$ if $r^n = x$. When n is even, then there are two roots of x, r and $-r$, since both $r^n = x$ and $(-r)^n = x$. Most times, however, $\sqrt[n]{x}$ represents the principal, or positive, nth root of x. When no index is shown, the radical alone represents the square root of a number ($n = 2$).

▶ **Example**

What is $\sqrt{81}$?

Since $9^2 = 81$, $\sqrt{81} = 9$.

▶ **Example**

What is $\sqrt[3]{729}$?

Since $9^3 = 729$, $\sqrt[3]{729} = 9$.

If $\sqrt[n]{x} = r$ where r is an integer, then x is called a perfect nth power. From the examples above, 81 is a perfect second power, or perfect square, and 729 is a perfect third power, or perfect cube.

Using Roots to Solve Equations

You can use roots to solve equations where a variable is raised to a power. Since taking a root is the inverse of taking a power, you can simplify these types of equations by taking the root of both sides.

▶ **Example**

Solve $x^2 = 16$.
By taking the square root of both sides, you can solve for x.

$$\sqrt{x^2} = \sqrt{16}$$
$$x = 4$$

Example

Solve $x^3 = 125$.

By taking the cube (3rd) root of both sides, you can solve for x.

$$\sqrt[3]{x^3} = \sqrt[3]{125}$$
$$x = 5$$

 TIP: If a number isn't a perfect nth power, you may have to approximate the solution. When this happens, the solution is an irrational number. For example, $\sqrt{2}$ is the irrational solution to $x^2 = 2$.

Practice

Directions: For questions 1 through 8, find the square root or cube root of each number.

1. $\sqrt[3]{343}$ _____

2. $\sqrt{121}$ _____

3. $\sqrt[3]{125}$ _____

4. $\sqrt[3]{512}$ _____

5. $\sqrt{169}$ _____

6. $\sqrt{625}$ _____

7. $\sqrt[3]{216}$ _____

8. $\sqrt{961}$ _____

Directions: For questions 9 through 14, solve the equation. For your solutions, use the principle root.

9. $x^2 = 36$ _____

10. $x^2 = 25$ _____

11. $x^3 = 27$ _____

12. $x^2 = 100$ _____

13. $x^3 = 64$ _____

14. $x^3 = 1,000$ _____

CCSS: 8.EE.6

Lesson 10: Graphing Equations of Lines

To find solutions to a linear equation in two variables, find a pair of numbers (x and y) that, when substituted into the equation, make the equation true. These values of x and y form an ordered pair (x, y). To graph a linear equation, use a table of values to generate at least three ordered pairs by choosing any arbitrary value for x or y, substituting that value into the equation, and then solving for the other variable. The easiest values to work with are usually $x = 0$ and $y = 0$.

Once you have found three ordered pairs, you can graph the solution by plotting them on a coordinate grid. If you have correctly found three ordered pairs, you can connect them to form a straight line.

 Example

Find three ordered pairs that are solutions to the following linear equation.

$x + 4y = 8$

Substitute $x = 0$, $y = 0$, and $x = 4$ into the equation to find three ordered pairs.

$(x = 0)$	$(y = 0)$	$(x = 4)$
$x + 4y = 8$	$x + 4y = 8$	$x + 4y = 8$
$0 + 4y = 8$	$x + 4(0) = 8$	$4 + 4y = 8$
$4y = 8$	$x + 0 = 8$	$4y = 4$
$y = 2$	$x = 8$	$y = 1$

Write these values in a table and find the ordered pairs.

x	y
0	2
8	0
4	1

Three ordered pairs that are solutions to the equation are (0, 2), (8, 0), and (4, 1).

Now that you have a few solutions to the linear equality, you can graph the solutions on a coordinate plane. Use the ordered pairs to plot the points on the graph, and then connect the points to form a line. The line represents all the solutions to the linear equation.

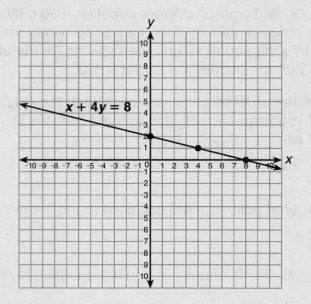

The ordered pairs (0, 2), (8, 0), and (4, 1) are only three solutions to the equation $x + 4y = 8$. There are an infinite number of solutions that are represented by the ordered pairs of every point on the line. In the example, the points (8, 0) and (0, 2) are called the intercepts of the graph. The **x-intercept** is the *x*-coordinate of the point (*x*, 0), where a line crosses (intercepts) the *x*-axis. Similarly, the **y-intercept** is the *y*-coordinate of the point (0, *y*), where a line crosses the *y*-axis.

 TIP: It is always a good idea to find at least three ordered pairs to make sure they all lie on a line. If they don't all lie on a line, then at least one of the ordered pairs is incorrect. Check your work.

CCSS: 8.EE.6

You can check to see if a point is a solution to a linear equation by substituting the *x*- and *y*-values into the equation.

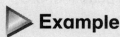 **Example**

Is (3, 4) a solution to the equation $3x - 2y = 1$?

Substitute 3 for *x* and 4 for *y* into the given equation.

$$3x - 2y = 1$$
$$3(3) - 2(4) \stackrel{?}{=} 1$$
$$9 - 8 \stackrel{?}{=} 1$$
$$1 = 1 \checkmark$$

Because the equation is true once you substitute these values for *x* and *y*, (3, 4) is a solution to $3x - 2y = 1$.

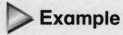 **Example**

Is (1, −1) a solution to the equation $5x + y = 7$?

Substitute 1 for *x* and −1 for *y* into the given equation.

$$5x + y = 7$$
$$5(1) + (-1) \stackrel{?}{=} 7$$
$$5 - 1 \stackrel{?}{=} 7$$
$$4 \neq 7$$

Because the equation is not true once you substitute these values for *x* and *y*, (1, −1) is not a solution to the equation $5x + y = 7$.

 TIP: You can use the graph of a linear equation to check if a point is a solution. If the ordered pair of the point lies on the graph of the equation, then it is a correct solution. If it does not lie on the graph of the equation, then it is not a correct solution.

53

Practice

Directions: For questions 1 through 4, fill in the table with three ordered pairs that are solutions to the linear equation, then graph the equation.

1. $5x + 4y = 20$

x	y

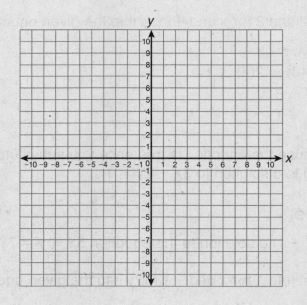

2. $3x + y = 5$

x	y

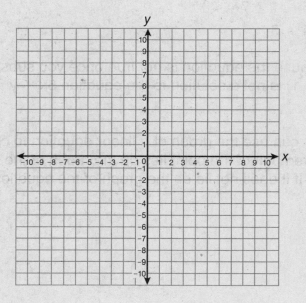

3. $6x - 2y = -4$

x	y

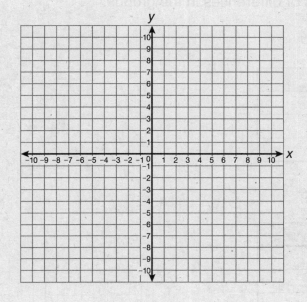

4. $-3x - y = 0$

x	y

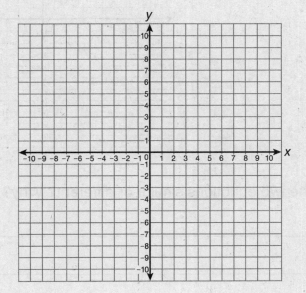

5. Graph the pairs of equations on the same coordinate plane, and then explain the similarities or differences in the graphs.

$4x + y = 6$

x	y

$2x + y = 6$

x	y

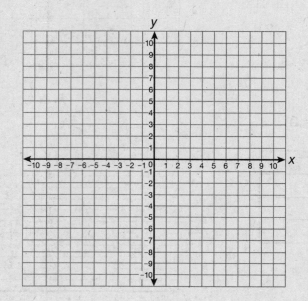

How are the two equations similar or different? Why?

_____

_____

CCSS: 8.EE.6

Lesson 11: Slope

Slope is the upward or downward slant of a line. A line that slants upward as you follow it from left to right has a positive slope. A line that slants downward as you follow it from left to right has a negative slope.

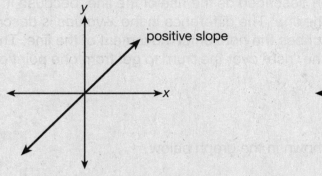

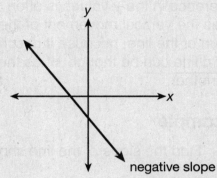

A horizontal line has a slope of zero. There is no vertical change in the line from left to right. A vertical line has an undefined slope.

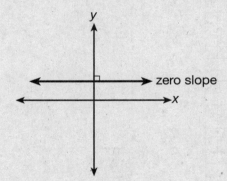

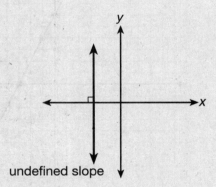

 TIP: A horiZontal line has a slope of Zero. An UnDefined slope goes Up and Down.

The slope of a line is the ratio of the vertical change of the line to its horizontal change. This ratio is a constant rate of change between any two points on the line. Slope can be determined using the following formula for two points on the line (x_1, y_1) and (x_2, y_2).

$$\text{slope} = \frac{y_2 - y_1}{x_2 - x_1}$$

The difference in the *y*-values is often described as the rise of the line, because it describes the vertical movement of the line. The difference in the *x*-values is described as the run of the line, because it describes the horizontal movement of the line. The slope of a line can be thought of as the "rise" over the "run" to get from one point on the line to another.

▶ Example

Find the slope of the line shown in the graph below.

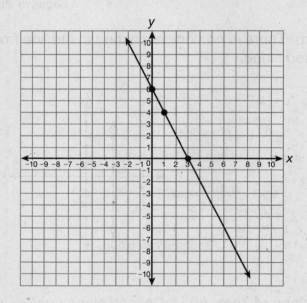

Choose two points on the line to represent (x_1, y_1) and (x_2, y_2).

(x_1, y_1): (0, 6) (x_2, y_2): (1, 4)

Substitute the values into the formula to find the slope.

$$\text{slope} = \frac{y_2 - y_1}{x_2 - x_1}$$

$$= \frac{4 - 6}{1 - 0}$$

$$= -\frac{2}{1} = -2$$

The slope of the line shown in the graph is −2.

CCSS: 8.EE.6

You can choose any two points to determine the slope of the line. The slope will be the same.

▷ Example

Three points on a line are (−8, 2), (2, −3), and (8, −6).

You can use any two points to determine the slope.

(x_1, y_1): (−8, 2) (x_2, y_2): (2, −3) (x_1, y_1): (2, −3) (x_2, y_2): (8, −6)

$$\text{slope} = \frac{y_2 - y_1}{x_2 - x_1} \qquad\qquad\qquad \text{slope} = \frac{y_2 - y_1}{x_2 - x_1}$$

$$= \frac{-3 - 2}{2 - (-8)} \qquad\qquad\qquad\qquad = \frac{-6 - (-3)}{8 - 2}$$

$$= -\frac{5}{10} \qquad\qquad\qquad\qquad\qquad = -\frac{3}{6}$$

$$= -\frac{1}{2} \qquad\qquad\qquad\qquad\qquad = -\frac{1}{2}$$

Both sets show that the slope of the line is $-\frac{1}{2}$.

If you plot the points (−8, 2), (2, −3), and (8, −6), you can see that they form similar triangles. The width of the smaller triangle is 6 units; its height is 3 units. The width of the larger triangle is 16 units; its height is 8 units. The sides of each triangle have the same ratio, meaning their third side has the same slope.

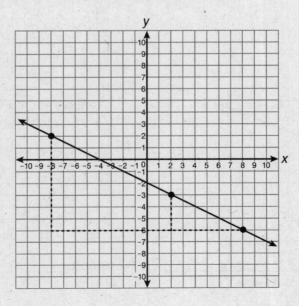

⬤ **Practice**

Directions: For questions 1 through 4, find the slope of the line shown in the graph.

1.

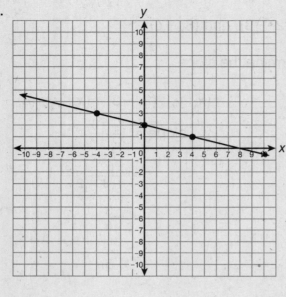

slope: _____

2.

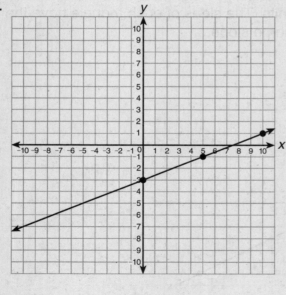

slope: _____

3.

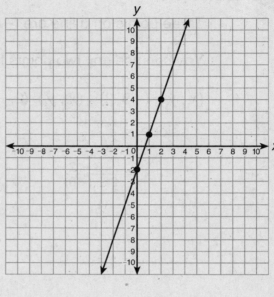

slope: _____

4.

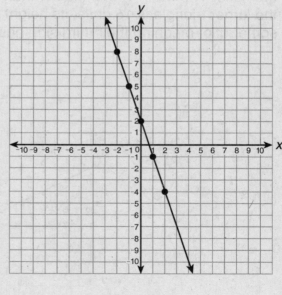

slope: _____

60

CCSS: 8.EE.6

5. What is the slope of the line shown on the following graph?

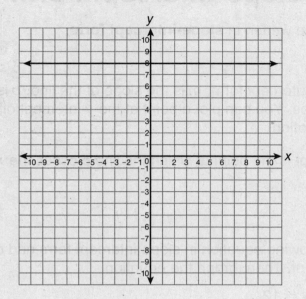

_____

Explain why you know the slope has that value. _____

_____

6. What is the slope of the line shown on the following graph?

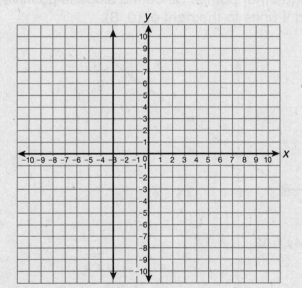

_____

Explain why you know the slope has that value. _____

_____

Lesson 12: Slope-Intercept Form

Linear equations can be written in **slope-intercept form**:

$$y = mx + b$$

When written in slope-intercept form, *m* is the slope of the line, *b* is the *y*-intercept, and *x* and *y* are the variables. You can plot a point at the *y*-intercept and then use the slope to graph the linear equation.

Likewise, the *x*-intercept is the point at which the line crosses the *x*-axis. The *y*-value at this point is always 0.

 Example

Write the following equation in slope-intercept form and determine the slope and *y*-intercept. Then, graph the equation.

$$-3x + 4y = 12$$
$$4y = 3x + 12$$
$$y = \frac{3}{4}x + 3$$

slope: $\frac{3}{4}$ *y*-intercept: 3

You can graph the equation by first plotting the *y*-intercept, (0, 3). Then, use the slope to find other points. Since the slope is positive, the next point is 3 units up and 4 units to the right of (0, 3).

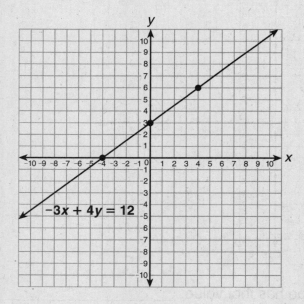

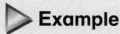

If an equation is in slope-intercept form, $y = mx + b$, you can determine the slope by finding the coefficient of x. You can think of lines that have a nonzero y-intercept ($b \neq 0$) as translations of lines that have the same slope, $y = mx$.

▷ Example

Graph $y = 3x$ and $y = 3x - 5$. Identify both of their slopes and y-intercepts. Find three points on each line and determine if there is a pattern in their y-values.

Both equations are in slope-intercept form.

$y = 3x$
slope: 3
y-intercept: 0

$y = 3x - 5$
slope: 3
y-intercept: -5

To graph both equations, start at their y-intercept, and find another point using the slope.

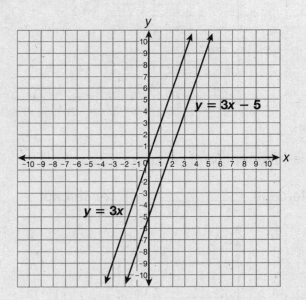

Three points that are on each line are shown in the table below.

x	$y = 3x$	$y = 3x - 5$
-1	-3	-8
0	0	-5
1	3	-2

Every y-coordinate of the points on the line $y = 3x - 5$ are 5 less than the y-coordinates of the points on the line $y = 3x$.

63

Practice

Directions: For questions 1 and 2, write the given equation in slope-intercept form. Then, find and use the slope, *y*-intercept, and *x*-intercept to graph the given linear equation.

1. $8x + 4y = 16$　　　　slope-intercept form _____

slope _____　　　　*y*-intercept _____　　　　*x*-intercept _____

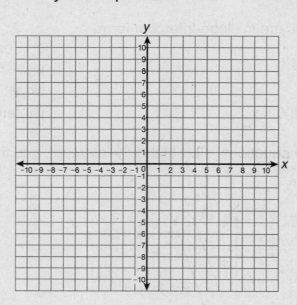

2. $7x - 3y = -12$　　　　slope-intercept form _____

slope _____　　　　*y*-intercept _____　　　　*x*-intercept _____

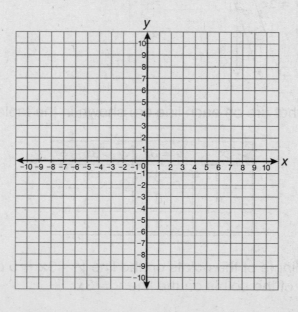

CCSS: 8.EE.6

Directions: For questions 3 through 6, write the given equation in slope-intercept form.

3. $2(2x - y) = 6$

   slope-intercept form _____

5. $-\frac{1}{4}x - y - \frac{1}{2}x = 5$

   slope-intercept form _____

4. $\frac{1}{3}y + x + \frac{1}{2}y = -3$

   slope-intercept form _____

6. $3(x + \frac{1}{9}y) = 1$

   slope-intercept form _____

7. Roberto needs to convert the equation $3x + y = -5$ into slope-intercept form. Why is he able to add or subtract the variable term from either side of the equation?

   _____

   _____

   _____

Lesson 13: Proportional Relationships

A **proportional relationship** shows how two variables may be related. When two variable quantities have a proportional relationship, one quantity changes when the other quantity changes. The changes occur in a predictable way so that the ratio of the variable quantities is always constant.

A proportional relationship can be expressed as $y = mx$, where m is the **unit rate** and describes how y changes per single unit of x. The unit rate, m, is the only constant part of the equation. To write a proportional relationship in this form, substitute the information you know.

 Example

Jon makes $8.50 per hour. Write a proportional relationship that shows how much Jon makes, y, based on the number of hours he works, x.

In this example, the unit rate is $8.50 per hour. The amount of money Jon makes and the number of hours he works are unknown quantities, so they can be left in variable form.

$$y = mx$$
$$y = (8.50)x$$

A proportional relationship that shows how much Jon makes is $y = 8.50x$. As x or y changes, the other variable will change proportionally.

In many proportional relationships, the unit rate, m, is not given. However, certain quantities may be known. An equation in the form $y = mx$ can be used to determine the unit rate.

 Example

A stationery store sells 3 ballpoint pens for $2.25. Write a proportional relationship that shows the cost of pens, y, based on the number of pens bought, x, and find the value of m.

The unit rate is unknown. The cost of the pens, y, is 2.25 and the number of pens is 3. Isolating m on one side of the equation provides the unit rate.

$$y = mx$$
$$(2.25) = m(3)$$
$$2.25 = 3m$$
$$0.75 = m$$

A proportional relationship that shows the cost of the pens is $2.25 = 3m$. The unit rate, m, is 0.75. This represents the price per pen, $0.75. You can now use that information to solve for the cost of any number of ballpoint pens.

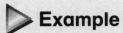CCSS: 8.EE.5

A proportional relationship in the form of the equation $y = mx$ can be illustrated with a graph. The slope of the equation represents the unit rate, m.

▷ **Example**

A county hospital wants to keep a proportion of nurses to doctors on its staff. There are currently 36 nurses and 9 doctors on staff. Create a graph that shows the proportional relationship of doctors to nurses in the hospital.

Use the equation $y = mx$ to first determine the unit rate, m.

$$y = mx$$
$$(9) = m(36)$$
$$9 = 36m$$
$$\frac{1}{4} = m$$

A proportional relationship that shows the relationship of doctors to nurses is $9 = 36m$. The unit rate, m, is $\frac{1}{4}$.

Graph a line that shows this relationship. The graph should go through point $(36, 9)$ with a slope of $\frac{1}{4}$.

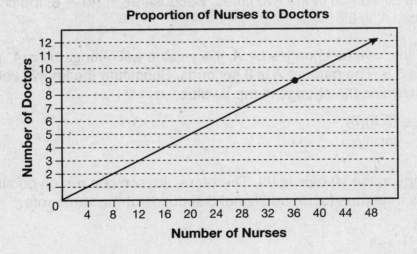

Proportion of Nurses to Doctors

The slope of the equation, m, is $\frac{1}{4}$. This slope represents the unit rate of doctors to nurses in the hospital.

When variable quantities are proportional to each other, a change in the *x*-value will change the *y*-value. If the *x*-value changes by an amount *A*, then the *y*-value changes by *mA*.

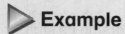 **Example**

Jayson ran 220 meters in 40 seconds. At that same rate, how much farther can Jayson run if he runs for 48 seconds?

The unit rate, *m*, of Jayson's speed can be determined using the equation *y* = *mx*, where *y* is the distance, in meters, and *x* is the time, in seconds.

$$y = mx$$
$$(220) = m(40)$$
$$220 = 40m$$
$$5.5 = m$$

The relationship of distance run to seconds is 220 = 40*m*. The unit rate, *m*, is 5.5 meters per second.

To solve this problem, you need to find the change in the *y*-value based on the change in the *x*-value. The change in the *x*-value can be determined by finding the difference in the two times. Because 48 − 40 = 8, the change in the *x*-value, *A*, is 8.

Because the *x*-value changes by *A*, the *y*-value will change by *mA*. The unit rate, *m*, is 5.5. The amount *A* is 8 seconds. Substitute the known values into *mA* to determine the change in the *y*-value.

$$mA = (5.5)(8)$$
$$= 44$$

The change in the *y*-value is 44. Therefore, Jayson can run an additional 44 meters if he runs for an additional 8 seconds at the same rate.

CCSS: 8.EE.5

 Practice

Directions: For questions 1 through 3, find an equation of the given proportional relationship.

1. An airplane flies at a constant speed and travels a distance of 360 miles in 45 minutes.

   _____

2. Emma answered 22 questions correctly on a multiple-choice test where each question was worth the same number of points. She scored an 88 on the test.

   _____

3. The Juarez household used 920 kilowatt-hours of electricity last month at a cost of $110.40.

   _____

Directions: For questions 4 through 6, use a proportional relationship to find the missing value.

4. Santiago rides his bicycle at a rate of 6 kilometers every 15 minutes. At that same rate, how many kilometers can Santiago ride in 25 minutes?

   _____

5. A photocopier makes 8 copies in 20 seconds. At that same rate, how many whole copies can the photocopier make in 48 seconds?

   _____

6. A laundromat charges $10.80 to clean 18 pounds of laundry. At that same rate, how much will the laundromat charge for 28 pounds of laundry?

   _____

Directions: For questions 7 and 8, graph the given proportional relationship, and then use the graph to find a given value.

7. Six ounces of silver is worth $120. Graph the relationship of the value of the silver to its weight.

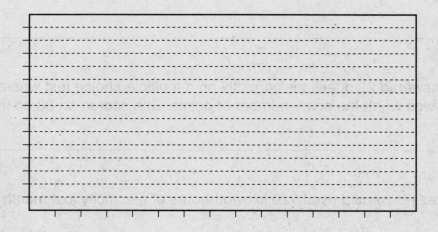

What is the weight of $400 worth of silver?

_____

8. A 16-ounce bag of lawn seed will cover a lawn that is 800 square feet. Graph the relationship of the size of a lawn to the number of ounces of lawn seed needed.

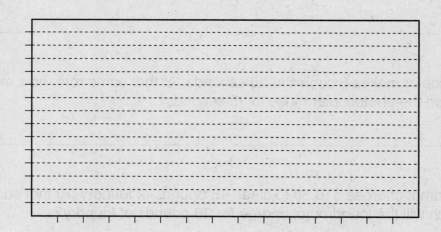

How many ounces of lawn seed are needed to cover a lawn that is 3,000 square feet?

_____

70

CCSS: 8.EE.5

Directions: For questions 9 through 12, find the change in the *x*- or *y*-value given the change in the other variable in the proportion to help solve for the unknown.

9. A photography website charges $18 for 6 small-size prints. How much more will it charge for 2 additional small-size prints?

_____

10. A bag with 8 cups of dried rice makes 32 servings of rice. How many fewer servings will the bag have if 2 cups are removed from the bag?

_____

11. A coffee shop charges a customer $10.50 for 1.5 pounds of organic coffee beans. How much more will the customer be charged if she wants an additional 0.5 pound of organic coffee beans?

_____

12. Drew puts 7 gallons of gasoline into his truck for a cost of $19.60. How much more will he have to pay if he adds 2 more gallons into the truck?

_____

Explain how you found the additional cost for Drew to add 2 more gallons into his truck.

_____

_____

_____

CCSS: 8.EE.5

Lesson 14: Comparing Proportions

Proportional relationships can be represented in different ways. For example, proportional relationships may be represented by an equation, a graph, or a table. You can compare the relationships of two or more proportions even though they may be represented differently. To do this, you need to determine the slope or unit rate of each proportional relationship. Then you can compare the slopes or unit rates.

 Example

Alexis has a job offer from two different companies. Company A offers a salary represented by the equation $y = 1200x$, where y is the total salary and x is the number of weeks of employment. Company B offers a salary represented by the following graph.

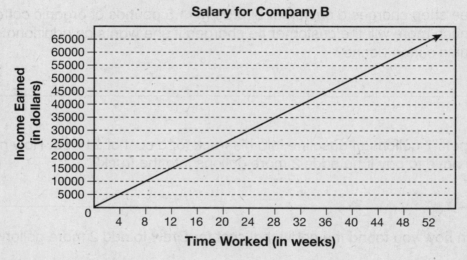

Which company offers the greater weekly salary?

The equation for the weekly salary of Company A is already in the form $y = mx$. Because *m* is the slope, the rate of change of the equation is $1200 for every week of employment. To determine the rate of change for the weekly salary for Company B, use the slope formula for any two points on the line. For example, use (4, 5000) and (20, 25,000):

$$\text{slope} = \frac{y_2 - y_1}{x_2 - x_1}$$

$$= \frac{25,000 - 5000}{20 - 4}$$

$$= \frac{20000}{16}$$

$$= 1250$$

The slope of Company B's graph is 1250. This means that the rate of change for Company B is $1250 for every week of employment. Because 1250 > 1200, Company B offers the greater weekly salary.

CCSS: 8.EE.5

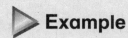 **Example**

The Cassini probe traveled through space toward Saturn at a constant speed represented by the following table.

Distance Traveled by Time of Cassini Probe

Distance Traveled (in km)	Elapsed Time (in seconds)
26 km	5 seconds
52 km	10 seconds
78 km	15 seconds
104 km	20 seconds

The New Horizons probe traveled through space toward Pluto at a constant speed represented by the following equation, where *y* is the distance traveled (in km) and *x* is the elapsed time (in seconds).

$$y = 19.79x$$

Which spacecraft traveled through space at a greater speed?

To solve this problem, you need to determine the speed for each spacecraft. You can do this by finding a common unit rate, kilometers per second. To find the unit rate for Cassini's speed, you can use the equation $y = mx$, where *y* is the distance traveled (in km) and *x* is the elapsed time (in seconds). Choose any point from the table, such as (52, 10).

$$y = mx$$
$$52 = 10m$$
$$5.2 = m$$

The unit rate for Cassini's speed is 5.2 kilometers per second.

To find the unit rate for New Horizons' speed, identify 19.79 as the unit rate 19.79 kilometers per second.

Because $19.79 > 5.2$, New Horizons traveled through space at a greater speed.

Practice

Directions: For questions 1 through 3, compare the proportional relationships to answer the questions.

1. The Sock Store sells pairs of socks according to the following graph.

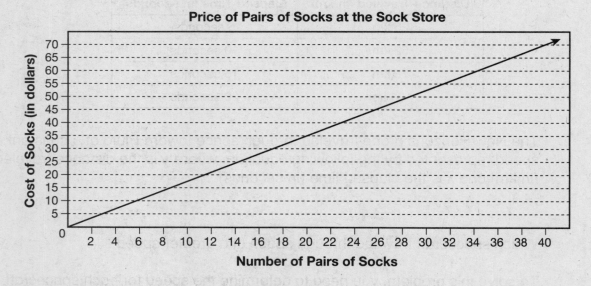

The Sock Stand sells socks according to the following equation, where y is the cost of the pairs of socks, in dollars, and x is the number of pairs of socks: $y = \frac{5}{3}x$.

Which seller offers the cheaper price for a pair of socks? _____

2. Two marathon runners pace themselves so that they each run at a steady speed for the entire race.

Ivan runs the marathon at a speed of 39 minutes for every 4 miles.

Meredith runs the marathon at a speed shown by the data in the following table.

Meredith's Marathon

Time Elapsed (in minutes)	Distance Run (in miles)
28 minutes	3 miles
56 minutes	6 miles
94 minutes	9 miles
112 minutes	12 miles

Both Ivan and Meredith finished the marathon. Who finished first?

3. The Health Food Store charges for granola according to the following graph.

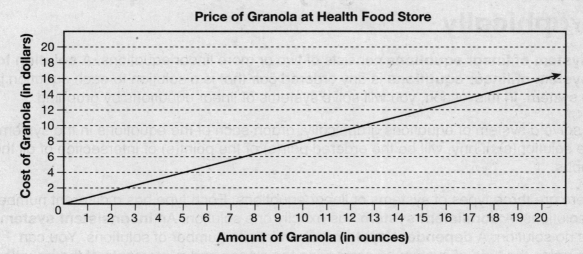

The Bulk Store sells granola in larger-quantity packages. The prices of its packages are shown in the following table.

Amount of Granola (in ounces)	Cost of Granola (in dollars)
16 ounces	$12
32 ounces	$24
48 ounces	$36
64 ounces	$48

Which store sells the granola for the cheaper per-ounce rate?

_____

Explain your reasoning or how you determined the answer.

_____

_____

_____

CCSS: 8.EE.8a, 8.EE.8b

Lesson 15: Solving Systems of Equations Graphically

A **system of linear equations** consists of two or more linear equations. A **solution to a system of linear equations** is any ordered pair that is a solution to each equation in the system. In this lesson, you will solve systems of linear equations by graphing.

To solve a system of equations graphically, graph each of the equations in the system. The solution(s), if any, will be the ordered pair(s) of the point(s) of intersection of all the graphs.

There are three types of systems of linear equations. Each type has a different number of solutions. A **consistent system** has exactly one solution. An **inconsistent system** has no solution. A **dependent system** has an infinite number of solutions. You can determine the type of system by comparing the slopes and *y*-intercepts of the equations in the system.

Consistent System

- exactly one solution
- different slopes

Inconsistent System

- no solution
- same slope, different *y*-intercepts

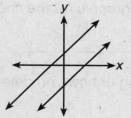

Dependent System

- infinite number of solutions
- same slope, same *y*-intercept

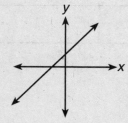

76

CCSS: 8.EE.8a, 8.EE.8b

Consistent System

A consistent system will have exactly one solution. The graphs of the lines in the system will intersect at one point. The solution will be the ordered pair of this point of intersection. In this type of system, the slopes of the graphs are different.

▶ Example

Solve the following system of equations graphically.

$$2x - y = 4$$
$$x + 4y = 20$$

Step 1: Write each equation in the system in slope-intercept form.

$$2x - y = 4$$
$$-y = -2x + 4$$
$$y = 2x - 4$$

slope: 2
y-intercept: -4

$$x + 4y = 20$$
$$4y = -x + 20$$
$$y = -\frac{1}{4}x + 5$$

slope: $-\frac{1}{4}$
y-intercept: 5

By looking at the slopes, you should be able to determine the type of system of equation that is graphed. Since the slopes are different, this system is consistent and the graph will show a pair of intersecting lines.

Step 2: **Plot each *y*-intercept. Then find other points using the slope. Draw the graph of each equation on the same coordinate plane.**

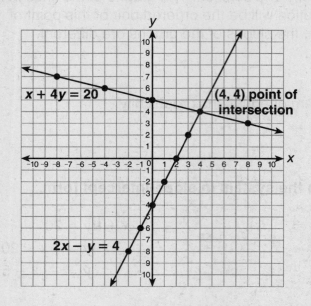

The two lines appear to intersect at (4, 4). This is a possible solution, but you need to check to see that it is a solution to both equations in the system.

Step 3: **Check the point of intersection.**

Substitute the values for the variables into each of the original equations to see if the ordered pair is a solution to both.

Check (4, 4): $2x - y = 4$ $x + 4y = 20$

$2(4) - (4) \stackrel{?}{=} 4$ $(4) + 4(4) \stackrel{?}{=} 20$

$8 - 4 \stackrel{?}{=} 4$ $4 + 16 \stackrel{?}{=} 20$

$4 = 4 ✓$ $20 = 20 ✓$

Since the ordered pair is a solution to both equations in the system, the solution to the system is (4, 4). This is a consistent system of linear equations.

78

CCSS: 8.EE.8a, 8.EE.8b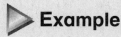

Inconsistent System

An inconsistent system will have no solution. The graphs of the lines in the system will be parallel. In this type of system, the slopes of the graphs in the system are the same and the *y*-intercepts of the graphs are different.

▷ **Example**

Solve the following system of equations graphically.

$$x + 4y = 12$$
$$3x + 12y = -24$$

Write each equation in the system in slope-intercept form. Then graph each of them on the same coordinate plane.

$x + 4y = 12$	$3x + 12y = -24$
$4y = -x + 12$	$12y = -3x - 24$
$y = -\frac{1}{4}x + 3$	$y = -\frac{1}{4}x - 2$
slope: $-\frac{1}{4}$	slope: $-\frac{1}{4}$
y-intercept: 3	*y*-intercept: -2

The slopes are the same and the *y*-intercepts are different, so the graph will show a pair of parallel lines.

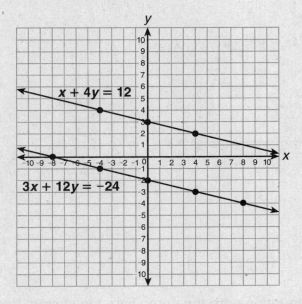

Since parallel lines do not intersect, the system has no solution. This is an inconsistent system of linear equations.

Dependent System

A dependent system will have an infinite number of solutions. The graphs of the lines in the system will be the same, so any point on the graph of one equation is also on the graph of the other equation. In this type of system, both the slopes and y-intercepts of the graphs are the same.

▷ Example

Solve the following system of equations graphically.

$4x - y = -9$
$-12x + 3y = 27$

Write each equation in the system in slope-intercept form. Then, graph each of them on the same coordinate plane.

$4x - y = -9$	$-12x + 3y = 27$
$-y = -4x - 9$	$3y = 12x + 27$
$y = 4x + 9$	$y = 4x + 9$
slope: 4	slope: 4
y-intercept: 9	y-intercept: 9

The slopes and y-intercepts are the same, so the graph will show one line.

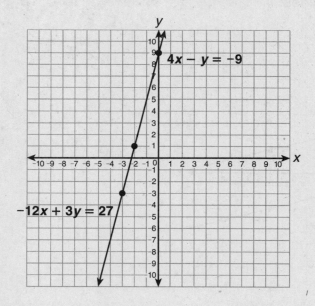

Since the graphs of the equations are the same line, the system has an infinite number of solutions. The solutions are all the ordered pairs that are solutions of either $4x - y = -9$ or $-12x + 3y = 27$. This is a dependent system of linear equations.

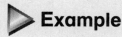
CCSS: 8.EE.8a, 8.EE.8b

Estimate Solutions by Graphing

Some solutions of a system of equations will not include integer values. In those cases, you can estimate the approximate solution by identifying a nearby point on the graph.

▷ Example

Solve the following system of equations graphically.

$$x + 2y = 1$$
$$-8x + 8y = -35$$

Write each equation in the system in slope-intercept form. Then, graph each of them on the same coordinate plane.

$x + 2y = 1$ $-8x + 8y = -35$

$\qquad 2y = -x + 1$ $8y = 8x - 35$

$\qquad\quad y = -\frac{1}{2}x + \frac{1}{2}$ $y = x - \frac{35}{8}$

slope: $-\frac{1}{2}$ slope: 1

y-intercept: $\frac{1}{2}$ y-intercept: $-4\frac{3}{8}$

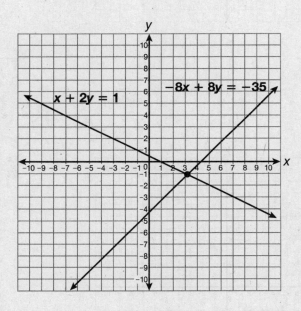

The solution to this system of equations does not include integer values. However, the solution is close to the point $(3, -1)$ on the graph. Therefore, you can estimate that the solution is about $(3, -1)$. This is a consistent system of linear equations.

CCSS: 8.EE.8a, 8.EE.8b

 Practice

Directions: For questions 1 through 4, solve the system of linear equations graphically. Then write the type of system for each.

1. $4x - y = -4$
   $12x - 3y = 24$

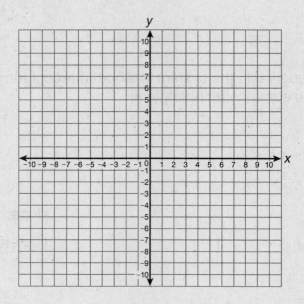

Solution(s): _____   _____

2. $3x - y = 7$
   $3x - 2y = 8$

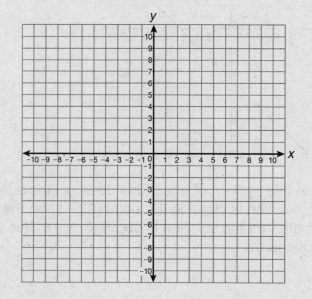

Solution(s): _____   _____

3. $2x + y = 7$
   $-3x + 3y = -6$

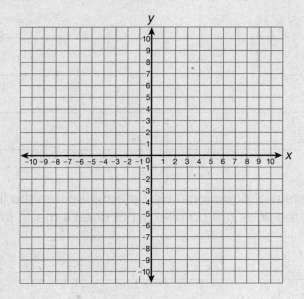

Solution(s): _____   _____

4. $-4x + 6y = -36$
   $2x - 3y = 18$

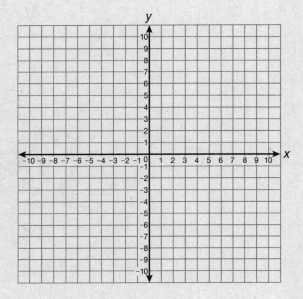

Solution(s): _____   _____

5. Karissa solves a system of equations graphically. Karissa's solution is (3, −5).

   Explain what her solution means to each equation in the system.

   _____

   _____

   _____

   _____

6. Aaron graphs a system of equations. He determines that the lines of the graph are parallel. Given that the lines are parallel, what does it mean for the system of equations?

   _____

   _____

   _____

   _____

CCSS: 8.EE.8b

Lesson 16: Solving Systems of Equations Using Substitution

One drawback to solving a system of equations graphically is that it can be hard to identify non-integer solutions with certainty. Or, a solution may be beyond the scope of the coordinate graph. Solving a system of equations using substitution will provide an exact solution. To solve a system of equations using substitution, you need to substitute the value of *x* or *y* in one equation for *x* or *y* in the other equation.

Consistent System

A consistent system has exactly one solution. That means the graphs of the equations would intersect at exactly one point.

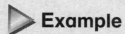 **Example**

Solve the following system of linear equations using substitution.

$$3x - 5y = -1$$
$$-3x + y = -3$$

Step 1: **Solve one of the equations for one of the variables.** In this case, solve the second equation for *y* because *y* already has a coefficient of 1.

$$-3x + y = -3$$
$$y = 3x - 3$$

Step 2: **Substitute the expression for the variable from Step 1 into the other equation and solve for the remaining variable.** Substitute $3x - 3$ for *y* in the first equation, and solve.

$$3x - 5y = -1$$
$$3x - 5(3x - 3) = -1$$
$$3x - 15x + 15 = -1$$
$$-12x = -16$$
$$x = \frac{4}{3}$$

Step 3: **Substitute the value for the variable into one of the original equations and solve for the remaining variable.** Substitute $\frac{4}{3}$ for *x* in the second equation and solve.

$$-3x + y = -3$$
$$-3\left(\frac{4}{3}\right) + y = -3$$
$$-4 + y = -3$$
$$y = 1$$

85

Step 4: **Check the values in the original equations.**

$$3x - 5y = -1 \qquad\qquad -3x + y = -3$$
$$3\left(\tfrac{4}{3}\right) - 5(1) \overset{?}{=} -1 \qquad\qquad -3\left(\tfrac{4}{3}\right) + 1 \overset{?}{=} -3$$
$$4 - 5 \overset{?}{=} -1 \qquad\qquad -4 + 1 \overset{?}{=} -3$$
$$-1 = -1 \checkmark \qquad\qquad -3 = -3 \checkmark$$

Since the ordered pair makes both equations true, $\left(\tfrac{4}{3}, 1\right)$ is the solution to the system. This is a consistent system of linear equations.

Inconsistent System

If the resulting equation is false, the system is inconsistent and has no solution. This means the graphs of the equations would be parallel.

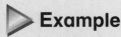 **Example**

Solve the following system of linear equations using substitution.

$$y + 4x = -14$$
$$2y + 8x = 18$$

In this case, solve the first equation for y because y already has a coefficient of 1.

$$y + 4x = -14$$
$$y = -4x - 14$$

Substitute $-4x - 14$ for y in the second equation and solve.

$$2y + 8x = 18$$
$$2(-4x - 14) + 8x = 18$$
$$-8x - 28 + 8x = 18$$
$$-28 \neq 18$$

Since the equation is false, there is no solution. This is an inconsistent system of linear equations.

CCSS: 8.EE.8b

Dependent System

If, when solving a system of equations using substituting, the resulting equation is true, the system is dependent and has an infinite number of solutions. The solution will be all the ordered pairs of all the points on the graph of either equation. That means the graphs of the equations would be the same.

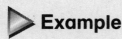 **Example**

Solve the following system of linear equations using substitution.

$$3y - x = -4$$
$$4x - 12y = 16$$

In this case, solve the first equation for x because x has a coefficient of -1.

$$3y - x = -4$$
$$-x = -3y - 4$$
$$x = 3y + 4$$

Substitute $3y + 4$ for x in the second equation and solve.

$$4x - 12y = 16$$
$$4(3y + 4) - 12y = 16$$
$$12y + 16 - 12y = 16$$
$$16 = 16 \checkmark$$

Since the equation is true, there are an infinite number of solutions. The solutions to the system are all the ordered pairs that make either $3y - x = -4$ or $4x - 12y = 16$ true. This is a dependent system of linear equations.

 TIP: You can use substitution to solve any system of linear equations. However, the easiest systems of equations to solve using substitution are those that have either the x or y variable in one of the equations with a coefficient of 1.

87

Practice

Directions: For questions 1 through 4, solve the system of linear equations using substitution. Then, write what kind of system it is.

1. $3x + 2y = 6$
   $x + 3y = -5$

   Solution(s): _____

   The system is _____.

3. $-x + 3y = 3$
   $2x - 6y = -6$

   Solution(s): _____

   The system is _____.

2. $y + 5x = -7$
   $2y + 10x = -8$

   Solution(s): _____

   The system is _____.

4. $2x + 4y = 3$
   $x - 2y = 1$

   Solution(s): _____

   The system is _____.

5. A line passes through points (0, 2) and (5, 7). Another line passes through points (−1, 6) and (3, 10). Determine whether the lines intersect.

Do the lines intersect?

_____

6. A line passes through points (−5, −2) and (−7, −6). Another line passes through points (1, 3) and (9, 7). Determine whether the lines intersect.

Do the lines intersect?

_____

7. Explain what it means when you arrive at a true statement when solving a system of equations using substitution.

_____

_____

8. Serena tries to solve the following system of equations using substitution.

$$8x + y = 10$$
$$2x + 3y = 8$$

Serena first isolates the y in the first equation to find that $y = 8x + 10$. She then substitutes $8x + 10$ for y in the second equation to get $2x + 3(8x + 10) = 8$. She then distributes to find $2x + 24x + 30 = 8$. Then she simplifies to find that $26x + 30 = 8$, or $26x = -22$. When she finds that $x = -\frac{11}{13}$, her teacher points out that she made a mistake. What mistake did Serena make?

_____

_____

_____

_____

Lesson 17: Solving Systems of Equations Using Linear Combinations

As with substitution, solving a system of equations using linear combinations will give you a precise solution. To solve a system of equations using linear combinations, multiply one or both equations by a constant factor, if necessary, so that the coefficients of either *x* or *y* are additive inverses (opposites). Then add the equations so that one variable adds to zero. The resulting equation can be solved to find the value of the other variable. Finally, substitute that value into one of the original equations and solve for the other variable.

Consistent System

A consistent system has exactly one solution.

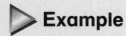 **Example**

Solve the following system of linear equations using linear combinations.

$$2x - 6y = 2$$
$$-4x + 8y = -8$$

The coefficients of neither *x* nor *y* are additive inverses. To change the coefficients of *x* so that they are additive inverses, multiply both sides of the first equation by 2. Then add the equations.

$$2(2x - 6y = 2) \rightarrow \quad 4x - 12y = 4$$
$$\underline{+ \ -4x + 8y = -8}$$
$$-4y = -4$$
$$y = 1$$

Now substitute *y* = 1 into either original equation and solve.

$$2x - 6y = 2$$
$$2x - 6(1) = 2$$
$$2x - 6 = 2$$
$$2x = 8$$
$$x = 4$$

Check the solution (4, 1) by making sure the values make both equations true.

$2x - 6y = 2$	$-4x + 8y = -8$
$2(4) - 6(1) \overset{?}{=} 2$	$-4(4) + 8(1) \overset{?}{=} -8$
$8 - 6 \overset{?}{=} 2$	$-16 + 8 \overset{?}{=} -8$
$2 = 2 \ ✓$	$-8 = -8 \ ✓$

The solution to the system is (4, 1). This is a consistent system.

CCSS: 8.EE.8b

Inconsistent System

If the resulting equation is false, the system is inconsistent and has no solution.

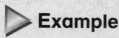 **Example**

Solve the following system of linear equations using linear combinations.

$$5y = 3x + 9$$
$$6x - 10y = -12$$

First, change the first equation so that it is in standard form: $Ax + By = C$.

$$5y = 3x + 9 \rightarrow -3x + 5y = 9$$

Now the two equations in the system are as follows:

$$-3x + 5y = 9$$
$$6x - 10y = -12$$

If you multiply the first equation by 2, the coefficient for x will become -6. That is the additive inverse of 6, the coefficient for x in the second equation. Multiplying the first equation by 2 will also change the coefficient for y to 10, the additive inverse of -10, the coefficient for y in the second equation.

$$-3x + 5y = 9 \rightarrow 2(-3x + 5y = 9) \rightarrow -6x + 10y = 18$$

Now you can add the two equations.

$$-6x + 10y = 18$$
$$\underline{+6x - 10y = -12}$$
$$0 \neq 6$$

Since the resulting equation is false, there is no solution. This is an inconsistent system.

 TIP: It is easiest to solve a system of equations using linear combinations when both equations are written in standard form: $Ax + By = C$. When substituting, always use an original equation from before you converted it into standard form.

Dependent System

If the resulting equation is true, the system is dependent and has an infinite number of solutions. The solution will be all the ordered pairs of all the points on the graph of either equation.

 Example

Solve the following system of linear equations using linear combinations.

$$-3x - y = 4$$
$$9x + 3y = -12$$

If you multiply the first equation by 3, the coefficient for y will become -3. That is the additive inverse of 3, the coefficient for y in the second equation.

$$-3x - y = 4 \rightarrow 3(-3x - y = 4) \rightarrow -9x - 3y = 12$$

Now you can add the equations.

$$-9x - 3y = 12$$
$$+9x + 3y = -12$$
$$\overline{ 0 = 0 \checkmark}$$

Since the resulting equation is true, there are an infinite number of solutions. This is a dependent system.

◯ Practice

Directions: For questions 1 through 4, solve the system of linear equations using linear combinations. Then, write what kind of system it is.

1. $5x - y = -6$
   $-2x + y = 3$

   Solution(s): _____

   The system is _____.

2. $2x + 6y = 6$
   $-x - 3y = -3$

   Solution(s): _____

   The system is _____.

3. $-2x + 2y = 4$
   $2x + 4y = 8$

   Solution(s): _____

   The system is _____.

4. $2x - y = 4$
   $y = 2x + 4$

   Solution(s): _____

   The system is _____.

5. Maximilian tried to solve the following system of linear equations using linear combinations.

$x - 4y = 7$
$3x + 2y = 7$

$3x + 2y = 7 \rightarrow 2(3x + 2y = 7)$
$2(3x + 2y = 7) \rightarrow 6x + 4y = 14$

$x - 4y = 7$
$6x + 4y = 14$
$7x = 7$
$x = 1$

$3x + 2y = 7$
$3(1) + 2y = 7$
$3 + 2y = 7$
$2y = 4$
$y = 2$

Maximilian's solution is (1, 2). The actual solution is (3, −1). What mistake did Maximilian make in his solution?

_____

_____

_____

_____

6. Explain what it means when you arrive at a false statement when solving a system of equations using linear combinations.

_____

_____

_____

_____

CCSS: 8.EE.8c

Lesson 18: Using Systems of Equations to Solve Problems

You can solve many real-world problems using systems of linear equations. To do this, you need to write a system of equations using the given information.

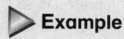 **Example**

Eve sells pretzels and juice at a pretzel stand. One morning she sells 14 pretzels and 8 juices and makes a total of $58. In the afternoon she sells 22 pretzels and 16 juices for $98. How much does Eve charge for one pretzel? How much does she charge for one juice?

To solve this problem, you need to set up linear equations to represent the information. Let $x =$ the cost of one pretzel and $y =$ the cost of one juice.

Eve sold 14 pretzels and 8 juices for $58. This can be represented by the equation $14x + 8y = 58$. Eve sold 22 pretzels and 16 juices for $98. This can be represented by the equation $22x + 16y = 98$.

You can use substitution or linear combinations to solve a system of equations. The coefficients for y are 8 and 16. You can multiply the first equation by -2 to change the coefficient to -16, the additive inverse of 16. Therefore, it may be easier to use linear combinations to solve.

$$14x + 8y = 58 \rightarrow -2(14x + 8y = 58) \rightarrow -28x - 16y = -116$$

Now you can add the equations.

$$
\begin{array}{r}
-28x - 16y = -116 \\
+22x + 16y = 98 \\
\hline
-6x = -18 \\
x = 3
\end{array}
$$

Substitute the value of x into one of the original equations to find the value of y.

$$14x + 8y = 58$$
$$14(3) + 8y = 58$$
$$42 + 8y = 58$$
$$8y = 16$$
$$y = 2$$

Check the solution (3, 2) by making sure the values make both equations true.

$$14x + 8y = 58 \qquad\qquad 22x + 16y = 98$$
$$14(3) + 8(2) \stackrel{?}{=} 58 \qquad\qquad 22(3) + 16(2) \stackrel{?}{=} 98$$
$$42 + 16 \stackrel{?}{=} 58 \qquad\qquad 66 + 32 \stackrel{?}{=} 98$$
$$58 = 58 \checkmark \qquad\qquad 98 = 98 \checkmark$$

The solution to the system is (3, 2). That means the cost of each pretzel is $3 and the cost of each juice is $2. This is a consistent system.

▷ Example

An electrician charges a service fee to show up at a customer's home. He then charges an additional hourly fee for each hour worked. He charges one customer $260 for working 4 hours at the customer's home. He charges another customer $150 for working 2 hours at the customer's home. How much does the electrician charge for his service fee? How much does he charge for his hourly fee?

To solve this problem, you need to set up linear equations to represent the information. Let x = the service fee and y = the hourly fee.

The electrician charged $260 for the service fee plus 4 hours. That is equal to $x + 4y = 260$. The electrician charged $150 for the service fee plus 2 hours. That is equal to $x + 2y = 150$.

You can use substitution or linear combinations to solve a system of equations. The coefficients for x are 1. Therefore, it may be easier to use substitution to solve.

$$x + 4y = 260$$
$$x = -4y + 260$$

Now you can substitute $-4y + 260$ for x in the second equation.

$$x + 2y = 150$$
$$(-4y + 260) + 2y = 150$$
$$-2y + 260 = 150$$
$$-2y = -110$$
$$y = 55$$

Substitute the value of *y* back into one of the original equations to find the value of *x*.

$x + 4y = 260$
$x + 4(55) = 260$
$x + 220 = 260$
$x = 40$

Check the solution (40, 55) by making sure the values make both equations true.

$x + 4y = 260$	$x + 2y = 150$
$(40) + 4(55) \stackrel{?}{=} 260$	$(40) + 2(55) \stackrel{?}{=} 150$
$40 + 220 \stackrel{?}{=} 260$	$40 + 110 \stackrel{?}{=} 150$
$260 = 260$ ✓	$150 = 150$ ✓

The solution to the system is (40, 55). That means the cost of the electrician's service fee is $40 and the cost of each hour of labor is $55. This is a consistent system.

● Practice

Directions: For questions 1 through 4, set up and solve a system of linear equations to solve each real-world scenario. Use either substitution or linear combinations.

1. Nicole has a total of 31 coins in a jar. There are only dimes and quarters in the jar. The value of the dimes and quarters in the jar is $5.50. How many dimes, *d*, does she have in the jar? How many quarters, *q*, does she have in the jar?

_____

_____

_____

2. A potter sells each vase for the same amount and each mug for the same amount. One day she sells 4 vases and 3 mugs for $80. Another day she sells 3 vases and 6 mugs for $90. How much does she charge for each vase, *v*? How much does she charge for each mug, *m*?

_____

_____

_____

3. A cable company charges a flat monthly rate for its customers, plus the same amount for each movie ordered on demand. One customer ordered 6 movies in a month and was charged a total of $104. Another customer ordered 2 movies in a month and was charged a total of $88. What is the cable company's monthly rate, *r*? What does the cable company charge for each movie ordered on demand, *m*?

_____

_____

_____

4. An office manager buys 2 office chairs and 4 file cabinets for $380. Next year she buys 4 office chairs and 6 file cabinets for $660. What is the cost of each office chair, *c*? What is the cost of each file cabinet, *f*?

_____

_____

_____

Explain how you found the cost of each chair and file cabinet.

_____

_____

_____

Unit 2 Practice Test

Read each question. Choose the correct answer.

1. If $\frac{2}{3}x - 2 = 6$, what is the value of x?

   A. 2

   B. 4

   C. 9

   D. 12

2. If $x^2 = 196$, what is the value of x?

   A. 4

   B. 7

   C. 14

   D. 392

3. A restaurant owner buys tables and chairs for her restaurant. Each table costs $140. She needs 4 chairs with each table. If she buys 14 tables, each with 4 chairs, and she spends a total of $4,200, what is the price of each chair?

   _____

4. What is 0.00000628 in scientific notation?

   A. $6.28 \cdot 10^{-5}$

   B. $6.28 \cdot 10^{-6}$

   C. $6.28 \cdot 10^{5}$

   D. $6.28 \cdot 10^{6}$

5. What are the solutions, if any, of the following equation?

$$18 - \frac{t}{20} = -9$$

   _____

6. If $5b - 25 = 25$, what is the value of b?

   A. There is no solution.

   B. 0

   C. 10

   D. 50

7. What are the solutions, if any, of the following equation?

   $$6y - 3 = -6y + 3$$

   A. There is no solution.

   B. There are an infinite number of solutions.

   C. 0.5

   D. 2

8. Gerald pays $160 for a hard drive for his computer. If the cost of each gigabyte of the hard drive is $0.20, how many gigabytes does the hard drive have?

   _____

9. What are the solutions, if any, of the following equation?

   $$8a - 1 = -2(-4a + 2)$$

   _____

10. Ray is 3 years older than half his sister's age. If Ray is 10 years old, how old is his sister?

   _____

11. A long-distance jogger runs 6 miles in 51 minutes. If the jogger continues to run at the same pace, how much additional time will he need to run 2 more miles?

A. 8.5 minutes

B. 16 minutes

C. 17 minutes

D. 68 minutes

12. What is the equation of a line that has a slope of $-\frac{1}{3}$ and passes through the point $(3, -6)$?

A. $y = -\frac{1}{3}x - 5$

B. $y = -\frac{1}{3}x - 6$

C. $y = -\frac{1}{3}x - 3$

D. $y = -6x - \frac{1}{3}$

13. Mr. Cohen can drive his hybrid car for 144 highway miles on 3 gallons of gas. At that same rate, how many miles can Mr. Cohen drive with 7 gallons of gas?

A. 0.146 mile

B. 48 miles

C. 192 miles

D. 336 miles

14. Using linear combinations, what is the solution(s) to the following system of linear equations, if there is one?

$$-3x + 4y = -1$$
$$4x - 4y = -4$$

A. No solution

B. $(-3, -2\frac{1}{2})$

C. $(-5, -4)$

D. $(-4, -5)$

15. What is the solution to the following system of linear equations, using substitution?

$$3x + y = -9$$
$$2x + 3y = -13$$

A. $(-3, -2)$

B. $(-3, 0)$

C. $(2, 3)$

D. $(-2, -3)$

16. Write the slope of a line that passes through (2, 1) and (4, −5). _____

17. Allison draws a line that is parallel to another line with a slope of $-\frac{1}{5}$. What is the slope of Allison's line? _____

18. A lumberyard charges $2.50 for an 8-foot piece of lumber. Write an equation to represent the proportional relationship that can be used to find the unit price. _____

19. What is the equation of a line that has a slope of 4 and passes through the point (−2, −9)? _____

20. Solve for k. If there is no solution, write "No solution." If there are an infinite number of solutions, write "Any value makes the equation true." _____

$$\tfrac{1}{2}(13k - 7) = \tfrac{3}{4}(6k + 2)$$

21. If $8p + 1 = 1$, what is the value of p? _____

22. Simplify:

$$2^4 \cdot 2^3 \cdot 2^{-9}$$

_____

23. Use a graph to determine the solution(s) for the following system of equations.

$$3x - 2y = -3$$
$$-2x + 4y = -6$$

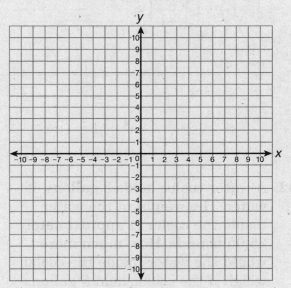

Solution(s): _____

24. Solve for d. If there is no solution, write "No solution." If there are an infinite number of solutions, write "Any value makes the equation true."

$$6(1.5d - 1) = 3(3d - 2)$$

_____

25. Yolanda gave 20% of her daily salary to taxes. She then put 10% of her remaining salary into her savings. If she is left with exactly $162, what is Yolanda's daily salary?

_____

26. If $7(t - 4) - 2t = 4(t - 3)$, what is the value of t?

_____

27. There are about $1.6 \cdot 10^5$ centimeters in 1 mile. How many centimeters are in $2.0 \cdot 10^3$ miles?

_____

28. Simplify:

$$\frac{5^3 \cdot 5^{-10}}{5^{-5}}$$

_____

29. Pablo graphs an equation in the form $y = mx + b$, as shown below.

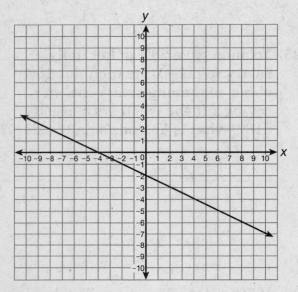

In Pablo's equation, $m = -\frac{1}{2}$. What is the value of b?

_____

30. A nursery charges $4.80 for 40 pounds of topsoil. Write an equation to represent the proportional relationship that can be used to find the unit price of the topsoil.

31. What is the slope of the line shown on the graph below?

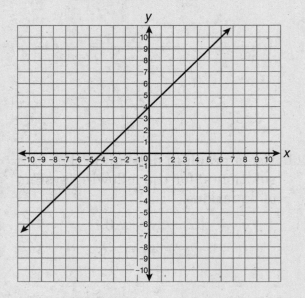

_____

32. A water cooler holds 3 gallons of water. The weight of the water inside the cooler is approximately 25.05 pounds. How much more would the water inside the cooler weigh if the cooler held 5 gallons instead of 3?

_____

33. What is the equation of a line that passes through point $(-3, -3)$ with a slope of $\frac{1}{2}$?

_____

34. Ms. Anastasia drives on the highway at a rate of 44 miles every 40 minutes. Write an equation to represent the proportional relationship that can be used to find the speed she travels, in miles per minute.

_____

35. If $50 = 3j - 23$, what is the value of j?

_____

36. A DVD burner adds \$105 to the price of a laptop computer. If the price of the laptop computer with the DVD burner is \$800, what is the price without the burner?

_____

37. What kind of system is represented by the following linear equations?

$$2x + 5y = -10$$
$$-4x - 10y = 20$$

_____

38. If $2l - 18 = 5l + 18$, what is the value of l? _____

39. Solve for q. If there is no solution, write "No solution." If there are an infinite number of solutions, write "Any value makes the equation true." _____

$$28.1q - 21.8 = 28.1q - 21.08$$

40. Solve for x. _____

$$x^3 = \frac{1}{1,000,000}$$

41. What is the equation of a line that passes through points $(0, 5.5)$ and $(3, -2)$? _____

42. A line passes through points $(-8, 1)$ and $(-6, 3)$. Another line passes through points $(1, 7)$ and $(-4, 2)$. Do the lines intersect? _____

43. What is the solution(s) to the following system of equations, if there are any?

$$x - 4y = 5$$
$$2x - 8y = -5$$

_____

44. What kind of system is represented by the following linear equations?

$$3x - y = -6$$
$$-x + y = 6$$

_____

45. The Mariana's Trench is the deepest trench in the ocean at 35,838 feet. There are about 30 centimeters in 1 foot. Estimate the depth of the Mariana's Trench in centimeters. Your answer should be written in scientific notation.

_____

46. Last weekend Lindsey bought 4 packs of hot dogs and 5 packs of hot dog buns for a barbeque. In total, she bought 62 hot dogs and hot dog buns. Each pack of hot dogs has the same number of hot dogs in it. Each pack of hot dog buns has the same number of buns in it. This weekend she bought 1 pack of hot dogs and 1 pack of hot dog buns for her family. In total, she bought 14 hot dogs and hot dog buns.

How many hot dogs come in each pack?

_____

How many hot dog buns come in each pack?

_____

47. Ryan's international phone service carrier charges a monthly rate of $17.95, plus $0.08 per minute. Sadie's international phone service carrier charges a monthly rate of $16.05, plus $0.09 per minute. If Sadie and Ryan paid the same amount last month for their international phone service and used the same number of minutes, how many minutes did Sadie use last month?

_____

Explain how you set up an equation to solve this word problem.

_____

_____

_____

_____

_____

48. Rasheed, a part-time clerk at a clothing store, gets paid with a constant hourly rate. One week he earns $135 for working 18 hours.

Part A
Graph the proportional relationship of the amount of money Rasheed earns to the number of hours he works.

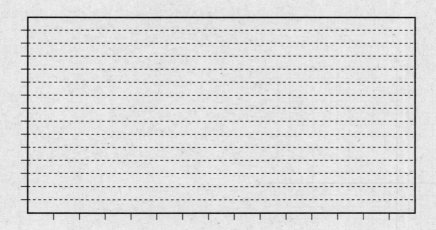

Part B
What is the slope of the proportional relationship of money earned to number of hours worked?

_____

Part C
Describe in words what the slope of the line means in the context of Rasheed's part-time job.

_____

_____

_____

_____

49. Erika needs to determine the solution(s) for the following system of equations.

$$6x + y = 4$$
$$8x - 2y = -4$$

Part A

Graph Erika's system of equations on the following coordinate graph. Then use the graph to estimate the solution.

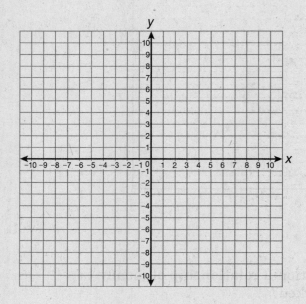

Estimated solution(s): _____

Part B

What kind of system is represented by Erika's system of linear equations?

_____

Part C

Use substitution or linear combinations to determine an exact solution for Erika's system of equations.

solution(s): _____

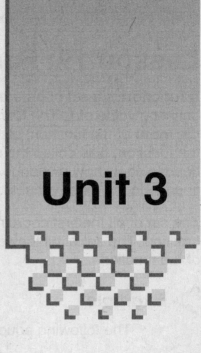

Unit 3

Functions

A function is one of the most important mathematical concepts in terms of its application in the real world. A function can be used to describe how changing one variable affects another variable. For example, functions are used to tell the boiling point of water at different altitudes, the speed of a skydiver at different times in a freefall, or the amount of money earned after different amounts of time.

In this unit, you will learn about linear and nonlinear functions. You will examine and determine the inputs and outputs of a function, called the domain and range. You will see all the different ways that a function can be represented—and then compare the functions in those different ways. You will write your own function to represent a real-life relationship and interpret the situation that it models. Finally, you will describe a function qualitatively by reading a graph or sketching it to show its features.

Lesson 19: Functions

A **function** is a set of ordered pairs (*x, y*) such that, for each value of *x*, there is one and only one value of *y*. The functions covered here are functions of *x*. This means that *x* is the **input** of the function, also called the **independent variable**, and *y* is the **output** of the function, also called the **dependent variable**. Equations, tables, and graphs can be used to show how a change in one variable in a functional relationship results in a change in the other variable.

The set of all the first coordinates, the input values, are called the **domain** of the function. The set of all the second coordinates, the output values, are called the **range** of the function.

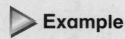 **Example**

The following equation represents a function.

$y = x - 3$

The equation can be described verbally.

y is equal to 3 less than *x*.

A table of ordered pairs can be made to represent the function.

x	y
−3	−6
0	−3
3	0

The ordered pairs from the table can be plotted on a coordinate plane. A line drawn through them shows the graph of the function.

In this example, a change in the value of *x* results in a change in the value of *y*.

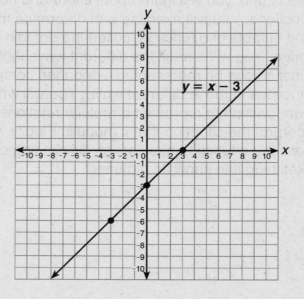

$y = x - 3$

112

CCSS: 8.F.1

You can evaluate a function for the output value, or range, if you are given the input value, or domain.

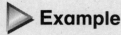 **Example**

Evaluate the following function for $x = 7$.
$$y = 2x - 1$$

To evaluate the function for a given value, substitute the known value into the function.
$$y = 2x - 1$$
$$y = 2(7) - 1$$
$$y = 14 - 1$$
$$y = 13$$

Therefore, when the input is 7 in the function $y = 2x - 1$, then the output is 13.

If you know the output value, or range, you can also use the function to determine the input value, or domain.

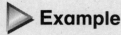 **Example**

Evaluate the following function if $y = 8$.
$$y = -3x + 2$$

To find the input of the function, plug the known value into the function.
$$y = -3x + 2$$
$$8 = -3x + 2$$
$$6 = -3x$$
$$-2 = x$$

Therefore, when the output is 8 in the function $y = -3x + 2$, then the input is -2.

Practice

Directions: For questions 1 through 7, determine whether the given table, graph, description, or equation represents a function.

1. $y = 7x + \frac{1}{8}$

   function? _____

2.

x	y
−5	5
4	4
5	5
8	8

   function? _____

3. *y* is equal to the quotient of 7 divided by *x*.

   function? _____

4.

x	y
−3	7
3	7
−2	7
−9	7

   function? _____

5.

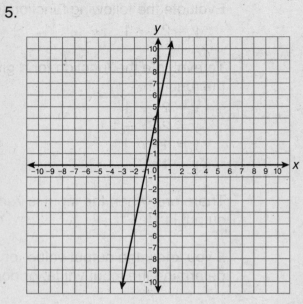

   function? _____

6. $x = 1$

   function? _____

7. *y* is multiplied by 10.

   function? _____

Directions: For questions 8 through 13, evaluate the function to determine the output.

8. For the function $y = -5x + 9$, the input is 4. What is the output?

    A. −29

    B. −11

    C.   1

    D.   11

9. For the function represented by the table below, what is the output when $x = 8$?

x	y
2	4
4	6
6	8
8	

    _____

10. y is equal to the product of 4 and x divided by 16; $x = -8$

    _____

11. $x = 0$ for the function $y = -12x + 88$

    _____

12. For the function graphed below, what is the output when $x = -3$?

    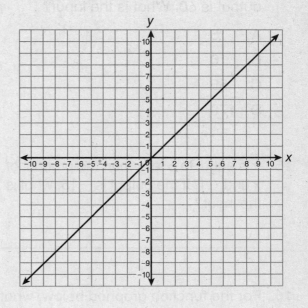

    _____

13. y is equal to the sum of 11 and x; $x = 39$

    _____

Directions: For questions 14 through 16, use the output of the function to determine the input.

14. In the function $y = 2x + 50$, the output is 60. What is the input?

    A.   5

    B.   10

    C.   55

    D. 170

15. y is equal to 6 less than the product of x and -1. If the output is -15, what is the input?

    _____

16. For the function graphed below, what is the output when $x = 2$?

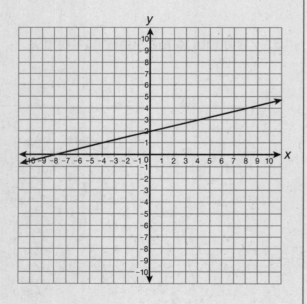

    _____

17. The following table represents a function. There is an input value missing from the table.

x	y
-10	-19
0	1
10	21
	41
40	81

    What value is missing from the table?

    _____

    Explain how you were able to determine the missing input value.

    _____

    _____

    _____

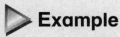

Lesson 20: Linear and Nonlinear Functions

When a function is represented by a graph, a line represents a linear function and a curve represents a nonlinear function.

▶ **Example**

The following graph represents a linear function.

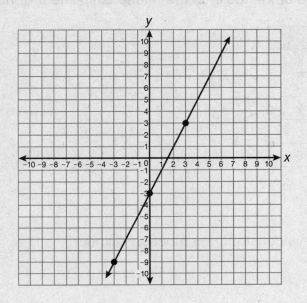

▶ **Example**

The following graph represents a nonlinear function.

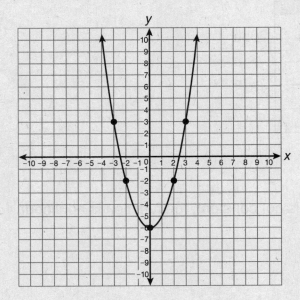

117

A linear function has a constant rate of change. A nonlinear function does not have a constant rate of change. A table can help show whether a function is linear or nonlinear. Examine the rate of change using the ordered pairs to see if there is a constant rate of change.

 Example

The following table represents a function with the rate of change between rows shown on each side of the table. Is the relationship linear or nonlinear?

x	y
−6	−1
−3	0
3	2
12	5
24	9

Left: 3, 6, 9, 12 Right: 1, 2, 3, 4

$$\frac{1}{3} = \frac{2}{6} = \frac{3}{9} = \frac{4}{12}$$

Since the rate of change is constant, the function is linear.

Example

The following table represents a function with the rate of change between rows shown on each side of the table. Is the relationship linear or nonlinear?

x	y
−1	−13
1	−3
5	5
8	14
9	18

Left: 2, 4, 3, 1 Right: 10, 8, 9, 4

$$\frac{10}{2} \neq \frac{8}{4} \neq \frac{9}{3} \neq \frac{4}{1}$$

Since the rate of change is not constant, the function is nonlinear.

118

CCSS: 8.F.3

If a function is linear, its equation can be written in **slope-interrept form** as $y = mx + b$, where m is the slope and b is the y-intercept. You can use this information to determine whether a function is linear or nonlinear from its equation.

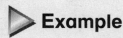 **Example**

Does the following equation represent a linear function?

$y = 3x - 8$

The given equation is in the form $y = mx + b$, so the function is a linear function. The graph of the function would be a straight line.

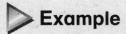 **Example**

Does the following equation represent a linear function?

$y = -2x^2 + 12$

The given equation is not in the form $y = mx + b$, so the function is not a linear function. The graph of the function would not be a straight line.

You can also write an equation to represent a linear function.

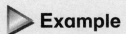 **Example**

The following table represents a linear function. What is its equation in slope-intercept form?

x	y
−2	8
0	2
2	−4
4	−10
6	−16

Each time the input value increases by 2, the output value decreases by 6. Therefore, the rate of change for the function is −3. When x is 0, y is 2. Therefore, the y-intercept is 2.

Now substitute the known values for the rate of change and the y-intercept into the form $y = mx + b$: $y = -3x + 2$.

⬤ Practice

Directions: For questions 1 and 2, determine whether the given graph or table represents a linear or nonlinear function.

1.

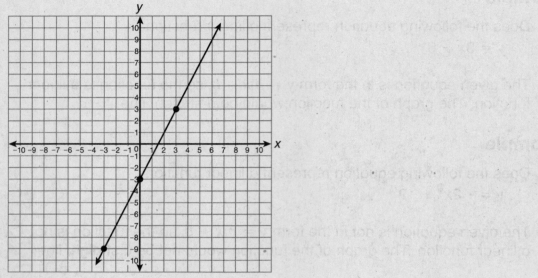

linear or nonlinear function? _____

2.

x	y
−4	8
−2	6
0	2
2	−2
4	−4

linear or nonlinear function? _____

CCSS: 8.F.3

Directions: For questions 3 and 4, determine whether the given table represents a linear function. If it does, write the equation of the function in slope-intercept form.

3.

x	y
−1	1
1	1
3	3
5	5
7	7

_____

4.

x	y
8	−3
16	−4
24	−5
32	−6
48	−8

_____

5. If the following table represents a linear function, write an equation in slope-intercept form to represent the function.

x	y
3	10
−1	−2
5	16
−2	−5
0	1

_____

Explain how you were able to identify whether the table represents a linear or nonlinear function.

_____

_____

_____

_____

Lesson 21: Comparing Functions

Two functions can be compared even if they are represented in different ways. To compare the functions, regardless of whether they are represented algebraically, graphically, verbally, or numerically in a table, determine the rate of change and *y*-intercept for each function. Then you can compare the functions' rates of change and their *y*-intercepts.

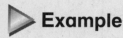 **Example**

Function A is represented by the following table.

x	y
−9	−10
−6	−6
−3	−2
3	6
6	10

Function B is represented by the following equation.

$$y = \frac{3}{4}x + 2$$

Compare functions A and B by their rates of change and their *y*-intercepts.

These functions are represented in different ways. To compare them, you need to find the rate of change and *y*-intercept of each function.

Each time the input value in function A increases by 3, the output value increases by 4. Because the function is linear, the output increases by $\frac{4}{3}$ when the input increases by 1. Therefore, the rate of change for function A is $\frac{4}{3}$. When *x* is 0, *y* is 2. Therefore, the *y*-intercept for function A is 2.

When a function is in slope-intercept form, $y = mx + b$, *m* represents the rate of change (or slope), and *b* is the *y*-intercept. Therefore, the rate of change for function B is $\frac{3}{4}$, and its *y*-intercept is 2.

Now that you know the rates of change and *y*-intercepts for both functions, you can compare them. Because $\frac{4}{3} > \frac{3}{4}$, the rate of change for function A is greater than the rate of change for function B. Because $2 = 2$, the *y*-intercepts for both functions are the same.

CCSS: 8.F.2

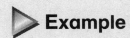 **Example**

Function C is represented by the following description.

The value of *y* is equal to the product of *x* and 2 plus 4.

Function D is represented by the following graph.

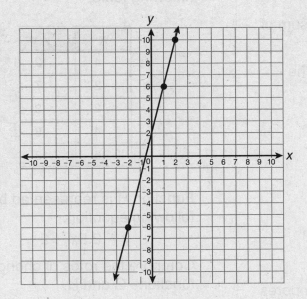

Compare functions C and D by their rates of change and their *y*-intercepts.

The description of function C can be translated into slope-intercept form as $y = 2x + 4$. Therefore, its rate of change is 2. Its *y*-intercept is 4.

You can find the rate of change of function D using the slope formula with any two points on the graph. Use (1, 6) and (2, 10), so $(x_1, y_1) = (1, 6)$ and $(x_2, y_2) = (2, 10)$.

Substitute the values into the formula to find the slope.

$$\text{slope} = \frac{y_2 - y_1}{x_2 - x_1}$$

$$= \frac{10 - 6}{2 - 1}$$

$$= \frac{4}{1}$$

$$= 4$$

The slope of the line in the graph is 4, so the rate of change of function D is 4. The line crosses the *y*-axis at 2, so the *y*-intercept of the function is 2.

Because $4 > 2$, the rate of change for function D is greater than the rate of change for function C. Because $2 < 4$, the *y*-intercept for function D is less than the *y*-intercept for function C.

123

⬤ Practice

Directions: For questions 1 through 3, compare the rates of change and the *y*-intercepts for the functions.

1. Function E is represented by the following description.

   The value of *y* is equal to the sum of *x* and −5.

   Function F is represented by the following table.

x	y
−5	0
−3	1
−1	2
1	3
3	4

   Which statement correctly compares the properties of functions E and F?

   A. Function E has a greater rate of change and a greater *y*-intercept than function F.

   B. Function E has a greater rate of change and a smaller *y*-intercept than function F.

   C. Function F has a greater rate of change and a greater *y*-intercept than function E.

   D. Function F has a greater rate of change and a smaller *y*-intercept than function E.

2. Function G is represented by the following table.

x	y
−4	12
−2	7
2	−3
4	−8
6	−13

   Function H is represented by the following equation.

   $$y = 2.5x - 2$$

   Compare the rates of change for functions G and H.

   _____

   _____

   Compare the *y*-intercepts for functions G and H.

   _____

   _____

124

CCSS: 8.F.2

3. Function J is represented by the following table.

x	y
−9	1
−3	3
3	5
6	6

Function K is represented by the following graph.

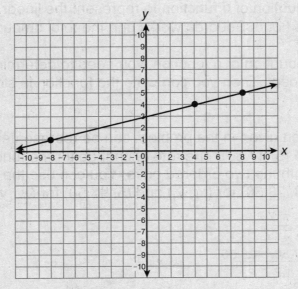

Compare the rates of change for functions J and K.

_____

Compare the *y*-intercepts for functions J and K.

_____

Explain how you were able to determine the rates of change for each function.

_____

_____

_____

Lesson 22: Applications of Functions

Functions can be used to model the relationship between two quantities in real-world problems. You can write a function to represent this relationship given two ordered pairs of the function.

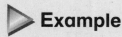 **Example**

When Eduardo works for 3 hours at his job, he earns a total of $28.50. When he works for 5 hours, he earns a total of $47.50.

Create the equation of a function to represent the linear relationship between the number of hours Eduardo works and the total amount of money he earns.

To write the equation of a function, you need to determine its rate of change and its y-intercept. Then you can write the function in slope-intercept form, $y = mx + b$.

The information given in the problem can be represented as ordered pairs: (3, 28.50) and (5, 47.50). The rate of change for the function can be determined with the two ordered pairs. You can use the slope formula, substituting (3, 28.50) for (x_1, y_1) and (5, 47.50) for (x_2, y_2).

$$\text{slope} = \frac{y_2 - y_1}{x_2 - x_1}$$
$$= \frac{47.50 - 28.50}{5 - 3}$$
$$= \frac{19}{2}$$
$$= 9.50$$

The rate of change is 9.50, which means that for every hour Eduardo works, he earns $9.50.

You can find the y-intercept of any linear function given its rate of change and one point on its line. Just substitute the known information into the slope-intercept form. Substitute the x- and y-coordinates from one point for x and y in the equation. Use $x = 3$ and $y = 28.50$.

$$y = mx + b$$
$$(28.50) = (9.50)(3) + b$$
$$28.50 = 28.50 + b$$
$$0 = b$$

Therefore, the y-intercept of the function is 0. Because its rate of change is 9.50, the function can be written as $y = 9.50x$. You don't need to include the y-intercept in the equation if it is equal to 0.

You can write a function to represent a real-world relationship given a linear relationship represented by a table. To do this, you need to determine the rate of change and *y*-intercept from the data in the table.

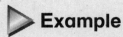 **Example**

A cell-phone company charges its customers different amounts according to customers' usage, as shown in the following table.

Hours of Usage, *x*	Total Charges (in $), *y*
3	37
5	45
7	53
10	65

Determine the hourly rate the cell-phone company charges its customers, and the cost of using 0 hours.

The rate of change for the linear function can be determined by two ordered pairs. You can use the slope formula to determine the rate of change, substituting the *x*- and *y*-values from any two points. Use (3, 37) for (x_1, y_1) and (10, 65) for (x_2, y_2).

$$\text{slope} = \frac{y_2 - y_1}{x_2 - x_1} = \frac{65 - 37}{10 - 3} = \frac{28}{7} = 4$$

The rate of change is 4, which means that the cell-phone company charges its customers $4 for every hour of cell-phone usage.

You can find the *y*-intercept of any linear function given its rate of change and one point on its line. Just plug the known information into the slope-intercept form. Substitute the *x*- and *y*-coordinates from one point for *x* and *y* in the equation. Use *x* = 3 and *y* = 37.

$y = mx + b$
$(37) = (4)(3) + b$
$37 = 12 + b$
$25 = b$

Therefore, the *y*-intercept of the function is 25. This means that the cell-phone company charges $25 if a customer uses 0 hours.

The function can be written as $y = 4x + 25$.

You can write a function to represent a real-world relationship given a graph. To do this, you need to determine the rate of change and *y*-intercept from the data shown in the graph.

 Example

Justine earns a different amount of money depending on how many hours she works at a car wash. The relationship between the money she earns and the hours she works is represented by the following graph. Find her hourly rate, and how much money she makes if she works no hours.

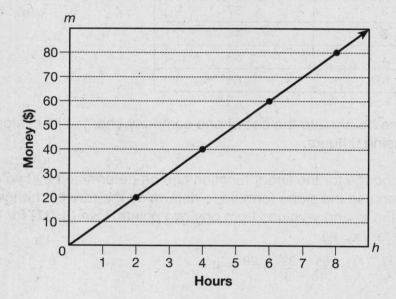

The rate of change for the linear function can be determined using the two points on the graph. You can use the slope formula, substituting the *x*- and *y*-values from any two points. Use (2, 20) for (x_1, y_1) and (4, 40) for (x_2, y_2).

$$\text{slope} = \frac{y_2 - y_1}{x_2 - x_1}$$

$$= \frac{40 - 20}{4 - 2}$$

$$= \frac{20}{2} = 10$$

The rate of change is 10, which means that Justine earns $10 for each hour worked at the car wash.

The *y*-intercept of the function is 0. Justine earns $0 if she works no hours.

The function can be written as $y = 10x$.

CCSS: 8.F.4

 Practice

Directions: For questions 1 through 3, determine the rate of change and *y*-intercept for each function.

1. A social networking site charges $40 for 2,000 page views of a banner advertisement. The site also charges $100 for 5,000 page views of a banner advertisement.

    Determine the rate of change for the function. _____

    Explain what the rate of change means in terms of this scenario.

    _____

    _____

    Determine the *y*-intercept for the function. _____

    Explain what the *y*-intercept means in terms of this scenario.

    _____

    _____

2. Shantell saves the same amount of money each month in her bank's savings account. The amounts of money she has saved after different numbers of months are shown in the following table.

Months of Saving, *x*	Total Amount Saved (in $), *y*
4	1100
6	1400
8	1700
10	2000

    Determine the rate of change for the function. _____

    Explain what the rate of change means in terms of this scenario.

    _____

    _____

    Determine the *y*-intercept for the function. _____

    Explain what the *y*-intercept means in terms of this scenario.

    _____

    _____

3. A shipping company uses the following graph to show its customers the cost of sending a package based on the weight of the package being sent.

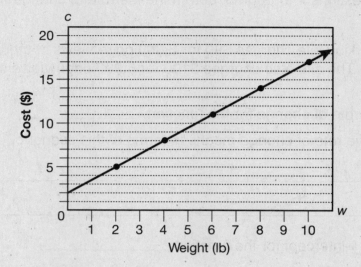

Determine the rate of change for the function. _____

Explain what the rate of change means in terms of this scenario.

_____

_____

Determine the *y*-intercept for the function. _____

Explain what the *y*-intercept means in terms of this scenario.

_____

_____

Explain why the graph of the function does not extend to the left side of the *y*-axis or below the *x*-axis.

_____

_____

130

Lesson 23: Qualitatively Describing Functions

Some functions can be described without specific values. This is called describing a function **qualitatively**, because only the qualities of the values matter—not the quantities. When a function is described qualitatively with a graph, it provides a way to identify basic relationships between the variables. For example, it can show whether a function is linear or nonlinear, or whether the function is decreasing or increasing.

When graphing a function qualitatively, the *x*-axis should be used to show the independent variable. The *y*-axis should be used to show the dependent variable.

 Example

A T-shirt printer charges her customers to print a specific pattern on a T-shirt. The printer charges less, per each T-shirt printed, as a customer orders a greater number of T-shirts. Create a graph to qualitatively describe the cost of printing T-shirts for a customer.

This problem does not provide numerical values or quantities, but the price of printing T-shirts can still be described qualitatively. To start, identify the two variables being modeled: the number of T-shirts printed and the total cost to make that number of T-shirts. The total cost to make a certain number of T-shirts depends on the number of T-shirts being made. The number of T-shirts is the independent variable, *x*, while the cost of that number of T-shirts is the dependent variable, *y*. Because there is only one cost to make *x* T-shirts, the relationship between *x* and *y* describes a function.

The cost of each T-shirt goes down with each T-shirt ordered, which means that the function is nonlinear.

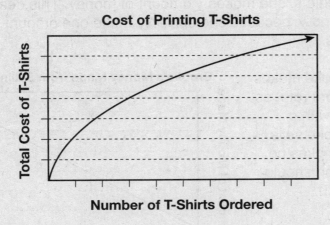

Cost of Printing T-Shirts

Total Cost of T-Shirts

Number of T-Shirts Ordered

▷ **Example**

A scuba diver has a certain amount of air in her tank at the start of a dive. The amount of air decreases at a constant rate, as long as she does not descend deeper than 10 meters below sea level. Create a graph to qualitatively describe the amount of air in the scuba diver's tank during a dive.

The amount of air left in the tank depends on the amount of time the diver has spent underwater. The independent variable, x, is the amount of time the diver has spend underwater. The dependent variable, y, is the amount of air left in her tank. At a specific time, x, there is only one amount of air in the diver's tank. This means that the relationship between x and y describes a function.

She starts with a certain amount of air, so the function has a positive y-intercept. The amount of air in the tank goes down at a constant rate, so the function is linear with a negative slope.

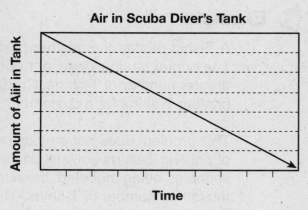

Air in Scuba Diver's Tank

▷ **Example**

Brooke sells each of her homemade bracelets on the Internet for the same amount. Create a graph to qualitatively describe how Brooke earns money from selling her bracelets.

If Brooke sells x bracelets, she makes y amount of money. This describes a function between x and y, because she can only make one amount of money for selling x bracelets.

The amount that Brooke makes increases at a constant rate, so the function is linear with a positive slope. If Brooke sells zero bracelets, she makes no money, so the function should have a y-intercept of 0.

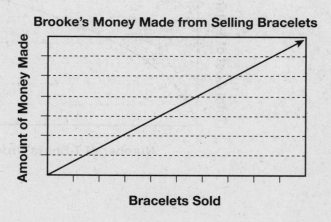

Brooke's Money Made from Selling Bracelets

CCSS: 8.F.5

⬤ Practice

Directions: For questions 1 through 7, graph the relationship between the two quantities. Label the axes for each graph.

1. A family drives a car at a constant speed for a certain number of minutes to reach a destination. Represent the time the family drives versus the distance the family travels.

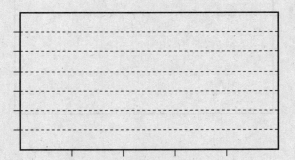

3. The more money that a company charges for an object, the fewer people that will buy it. The relationship changes at a constant rate. Represent the cost of an object versus the number of people that will buy it.

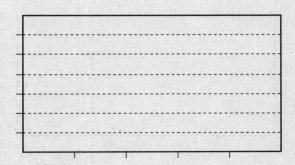

2. Javier will only get a few points on his social studies test if he does not study. He will get more points on the test for every hour that he studies. Represent the time Javier studies versus his points scored on the test.

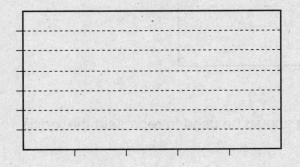

4. An Internet user pays a flat rate to be able to download an unlimited amount of data per month. Represent the amount of data downloaded versus the cost of the Internet service.

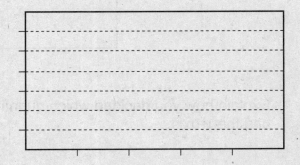

5. Yoko opens a high-yield savings account with a certain amount of money. Each day she earns interest on the money in her account, which then gets added back to the savings account. She does not add to or subtract any other money from the account. Represent the time Yoko has the account versus the amount of money in the account.

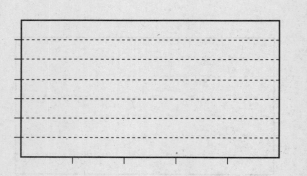

6. A submarine at the surface of the ocean has very little pressure pushing on its outer surface. As the submarine descends into the ocean, the pressure on its outer surface increases at a constant rate. Represent the depth of the submarine versus the pressure on its outer surface.

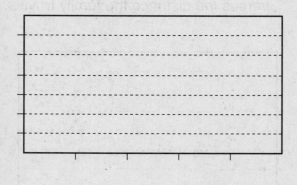

7. The Cohen family uses a gas-powered grill that runs on a tank of propane. The family begins the summer with a full tank of gas. They use the same amount of gas for each barbecue. Represent the number of times the family uses the grill versus the amount of gas left in the tank.

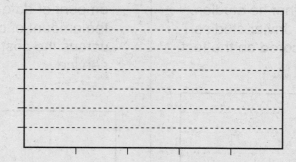

Explain how you decided which quantities should be used to represent the *x*-axis and the *y*-axis.

_____

_____

_____

134

Unit 3 Practice Test

Read each question. Choose the correct answer.

1. Which of the following tables does **not** represent a function?

   A.
x	y
−8	9
−4	9
0	9
4	9

   B.
x	y
3	5
1	3
−2	0
−9	−7

   C.
x	y
4	−1
1	−4
4	1
1	4

   D.
x	y
−10	12
−4	6
2	0
8	6

2. Function M is represented by the following description.

   The value of y is equal to 1 less than x.

   Function N is represented by the following graph.

   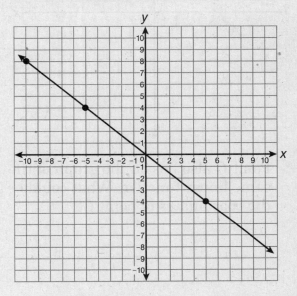

   Which statement correctly compares the properties of functions M and N?

   A. Function M has a greater rate of change and a greater y-intercept than function N.

   B. Function M has a greater rate of change and a smaller y-intercept than function N.

   C. Function N has a greater rate of change and a greater y-intercept than function M.

   D. Function N has a greater rate of change and a smaller y-intercept than function M.

3. Points *A* and *B* lie on the graph of a linear function, as shown below.

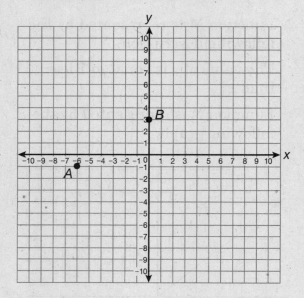

Point *C* also lies on the graph of the same linear function. What could be the coordinates of point *C*?

_____

4. In the function $y = 5x - 25$, the output is -25. What is the input?

_____

5. A hand sanitizer machine dispenses a standard amount of soap each time a button is pushed. The machine is filled to capacity and installed in a restroom. Create a graph to show the relationship between the number of times the button is pushed and the amount of soap left in the machine. Include titles for the axes.

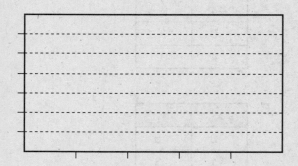

6. Evaluate the following function when the input is equal to 4.

   y is equal to the product of -3 and *x* added to 6.

_____

7. In the function $y = 24x$, the output is 8. What is the input?

_____

8. What is the range of the function represented by the following table?

x	y
−2	2
−1	1
1	1
2	2

_____

9. Ms. Wilkins draws the following table on the board in her class.

x	y
−9	8
−1	2
3	−1
7	−4
−5	5

Does Ms. Wilkins's table represent a linear or nonlinear function?

_____

10. The following graph represents the relationship between two quantities.

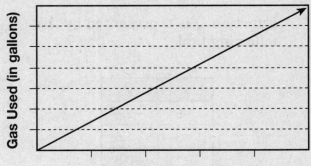

Distance Traveled (in miles)

Describe the relationship between the two quantities on the graph.

_____

_____

11. An input value is missing from the following table.

x	y
3	3
5	5
	7
9	9

Which input value would show that the table does **not** represent a function?

_____

12. Function P is represented by the following equation.

$$y = -2x - 5$$

Function Q is represented by the following table.

x	y
−5	−2.5
−3	−1.5
−1	−0.5
1	0.5

Which function has the greater rate of change?

_____

13. A renter pays her landlord the same amount every month to rent her apartment. Graph the relationship between the number of months and the amount paid per month. Include a title for each axis.

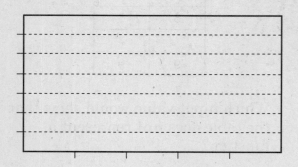

14. What is the domain of the function represented by the following table?

x	y
−8	−8
−5	−7
−2	−6
1	−5

_____

15. In the function $y = 6x + 4$, the output is 34. What is the input?

_____

16. Evaluate the following function when the input is equal to −8.

$$y = -2x - 6$$

_____

17. If the following table represents a linear function, write an equation in slope-intercept form to represent the function.

x	y
−8	3
−6	4
−4	5
−2	6
2	8

_____

18. Function R is represented by the following table.

x	y
−1	−10
1	−4
3	2
5	8

Function S is represented by the following description.

The value of y is equal to 2 times the value of x.

Which function has the greater y−intercept?

_____

19. In the function $y = -\frac{1}{2}x + 5$, if the output is $-\frac{1}{2}$, what is the input?

_____

20. Points D and E lie on the graph of a linear function, as shown below.

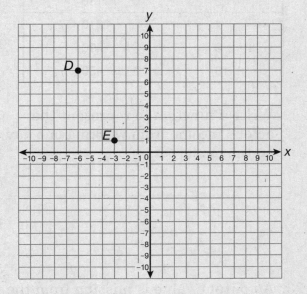

Point F also lies on the graph of the same linear function. What could be the coordinates of point F?

_____

21. What is the range of the following function if its domain is 4?

$$y = -5x + 18$$

_____

22. An actor gets paid to show up on a movie set. The actor is then paid an additional salary depending on the length of the workday, in hours. The producer of the film uses the following table to represent the total money that the actor will earn.

Hours of Work, x	Total Money Earned (in $), y
2	130
4	210
6	290
10	450

Write a function to represent the actor's income from working one day on the movie set.

_____

23. An output value is missing from the following table.

x	y
−9	5
−5	
3	−4
7	−7

What must the output value be for the table to represent a linear function?

_____

24. The following graph represents the relationship between the population of a species of birds and the time, in years.

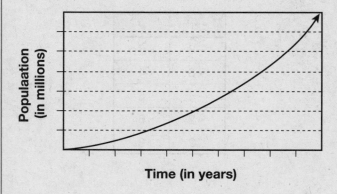

Describe the relationship between the two quantities on the graph.

_____

_____

_____

25. Translate the following description of a function into an equation in slope-intercept form.

y is equal to the product of −2 and x added to −5.

_____

140

26. The following graph shows four points of a function.

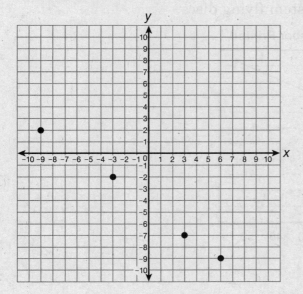

Do the ordered pairs of the function represent a linear or nonlinear function?

_____

Explain how you know whether the function is linear or nonlinear.

_____

_____

_____

27. Does the following table represent a function?

x	y
0	0
4	4
−4	4
7	7

_____

Explain how you know whether the table does or does not represent a function.

_____

_____

28. A sporting goods company makes custom flying discs for its customers. The company uses the following graph to show its customers the cost of printing custom flying discs.

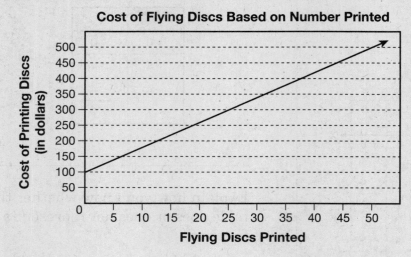

Cost of Flying Discs Based on Number Printed

Part A
Determine the rate of change for the function.

_____

Explain what the rate of change means in terms of this scenario.

_____

_____

Part B
Determine the y-intercept for the function. _____

Explain what the y-intercept means in terms of this scenario.

_____

_____

142

29. Team A will pay a professional athlete a signing bonus and then an annual salary. After two years, the athlete will have earned $280,000 playing for Team A. After four years, the athlete will have earned $440,000.

Team B will pay the same professional athlete according to a function represented by the following graph.

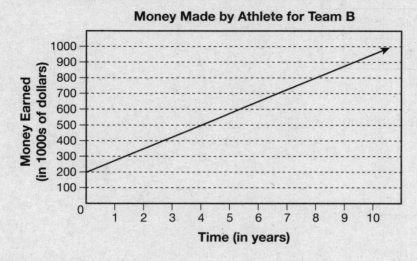

Part A

Which team offers the professional athlete the salary with the greater *y*-intercept? _____

Explain what the *y*-intercept means in terms of this scenario.

_____

_____

Part B

Which team offers the professional athlete the salary with the greater rate of change? _____

Explain what the rate of change means in terms of this scenario.

_____

_____

Part C

The professional athlete expects to play for exactly 10 years. Using 10 as the input, what will be the output, or total salary, for the functions used by each team?

Team A: _____   Team B: _____

30. A function is described in the table below.

x	y
−3	$-4\frac{3}{8}$
−1	$-1\frac{3}{8}$
4	$6\frac{1}{8}$
6	$9\frac{1}{8}$

Part A
Find the following of the function:

Slope: _____

y-intercept: _____

Equation: _____

Part B
If x changes by 20 units, what is the corresponding change in y?

_____

Explain how you arrived at your answer above.

_____

_____

Part C
Define another function by the sentence "The value of y is equal to one-third the value of x plus three-fifths." Which function has the steeper slope, and why?

_____

_____

_____

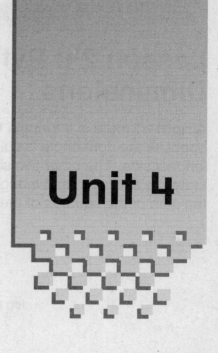

Unit 4

Geometry

Geometry is a powerful tool to describe the world around you. In fact, the prefix *geo-* means "Earth" and the root *-metry* means "to measure." Therefore, geometry is literally the measurement of the Earth. But geometry doesn't have to involve only physical objects. It can also relate to the shapes and figures on a flat surface. This is called two-dimensional geometry.

In this unit, you will examine the Pythagorean theorem and identify a proof for it. You will find the volumes of cones, cylinders, and spheres. You will transform figures using dilations, translations, rotations, and reflections. These transformations will be used to make predictions about the shape and size of a figure. You will also examine congruent and similar figures. Finally, you will draw and construct geometric figures.

In This Unit

145

CCSS: 8.G.7

Lesson 24: Pythagorean Theorem in Two Dimensions

A **right triangle** is a triangle that has one right angle. In a right triangle, the side opposite the right angle is called the **hypotenuse**. The hypotenuse is always the longest side. The other two shorter sides are called **legs**. The **Pythagorean theorem** states that for any right triangle, the square of the length of the hypotenuse is equal to the sum of the squares of the lengths of the legs.

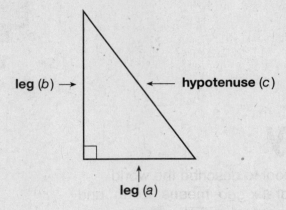

leg (b) →   ← hypotenuse (c)

↑
leg (a)

The Pythagorean theorem can be represented by the equation: $a^2 + b^2 = c^2$, where a and b are the lengths of the legs of the triangle, and c is the length of the hypotenuse. The example below uses sides of squares to show the Pythagorean theorem.

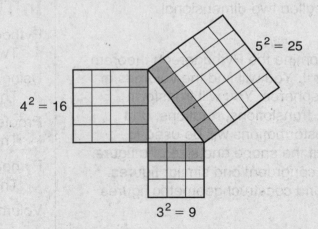

$5^2 = 25$

$4^2 = 16$

$3^2 = 9$

$a^2 + b^2 = c^2$

$(3)^2 + (4)^2 = (5)^2$

$9 + 16 = 25$

$25 = 25$ ✓

You can use the Pythagorean theorem to determine if a triangle is a right triangle. You can substitute the known side lengths into the Pythagorean theorem. If the equation is true, the triangle is a right triangle. If the equation is false, the triangle is not a right triangle.

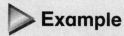 **Example**

Is the following a right triangle?

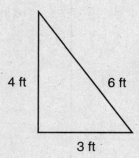

4 ft 6 ft 3 ft

Substitute the values into the Pythagorean theorem.

$a^2 + b^2 = c^2$

$(3)^2 + (4)^2 \stackrel{?}{=} (6)^2$

$9 + 16 \stackrel{?}{=} 36$

$25 \neq 36$

Because the equation is false, the triangle is not a right triangle.

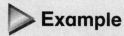 **Example**

Is the following a right triangle?

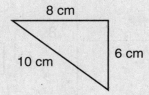

8 cm 6 cm 10 cm

Substitute the values into the Pythagorean theorem.

$a^2 + b^2 = c^2$

$(6)^2 + (8)^2 \stackrel{?}{=} (10)^2$

$36 + 64 \stackrel{?}{=} 100$

$100 = 100 \checkmark$

Because the equation is true, the triangle is a right triangle.

CCSS: 8.G.7

Practice

Directions: For questions 1 through 8, determine whether the triangle is a right triangle. Write "yes" or "no" on the line.

1.

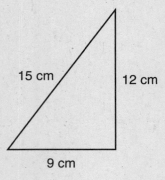

15 cm   12 cm

9 cm

_____

2.

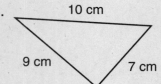

10 cm

9 cm   7 cm

_____

3.

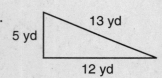

13 yd

5 yd

12 yd

_____

4.

12 ft

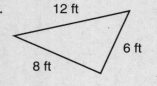

6 ft

8 ft

_____

5.

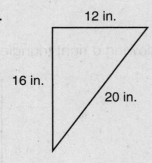

12 in.

16 in.   20 in.

_____

6.

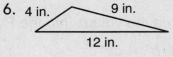

4 in.   9 in.

12 in.

_____

7.

8 m   13 m

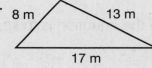

17 m

_____

8.

24 in.

7 in.

25 in.

_____

CCSS: 8.G.7

Lesson 25: Using the Pythagorean Theorem

You can use the Pythagorean theorem to find missing side lengths of right triangles. If you know that a triangle is a right triangle, you can substitute the known side lengths into the Pythagorean theorem and solve for the missing length.

 Example

What is the value of b in the following right triangle?

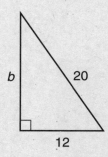

Substitute the known values into the Pythagorean theorem.

$$a^2 + b^2 = c^2$$
$$(12)^2 + b^2 = (20)^2$$
$$144 + b^2 = 400$$
$$b^2 = 400 - 144$$
$$b^2 = 256$$
$$b = \sqrt{256}$$
$$b = 16$$

The value of b is 16.

 TIP: Regardless of what letter the variable is, use the length of the hypotenuse for c and the length of the legs of the triangle for a and b.

CCSS: 8.G.7

The Pythagorean theorem can be applied to many real-world situations.

▷ **Example**

Omar drove a remote control boat across a creek that has a width of 9 ft. He stuck a flag in the creek bed to mark the starting point. When the boat got to the other side, the boat didn't cross the creek directly because the current carried the boat downriver. He wants to determine how far down the creek the boat ended up. If the boat actually traveled 15 ft, how far downriver did the boat end up?

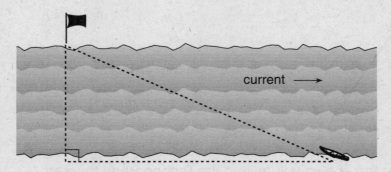

The width of the creek represents one leg of the right triangle. The distance the boat traveled represents the hypotenuse of the right triangle. The distance the boat ended up down the creek represents the other leg of the right triangle. To find this distance, substitute the known values into the Pythagorean theorem.

$$a^2 + b^2 = c^2$$
$$a^2 + (9)^2 = (15)^2$$
$$a^2 + 81 = 225$$
$$a^2 = 144$$
$$a = \sqrt{144}$$
$$a = 12$$

Omar's remote control boat ended up approximately 12 ft down the creek.

Practice

Directions: For questions 1 through 3, find the value of *x*.

1.

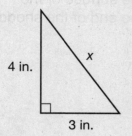

4 in.

x

3 in.

x = _____

2.

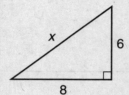

x

6

8

x = _____

3.

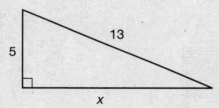

13

5

x

x = _____

4. What is the length of the hypotenuse of a right triangle that has one leg with a length of 9 inches and the other leg with a length of 12 inches?

   hypotenuse = _____

5. If the length of the hypotenuse of a right triangle is 10 feet and one leg measures 6 feet, what is the measure of the other leg?

   A. 4 feet

   B. 7 feet

   C. 8 feet

   D. 14 feet

Directions: For questions 6 through 9, use the Pythagorean theorem to find each missing value.

6. A lighthouse that is 120 ft tall casts a 160-ft long shadow on the surface of the water. What is the distance from the top of the lighthouse to the end of the shadow?

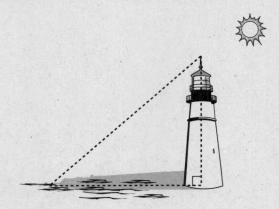

_____

7. Toby positioned a 13-ft ladder against the side of his house so he could paint. The distance from the base of the house to the top of the ladder is 12 ft. How far is the base of the ladder from the base of the house?

_____

CCSS: 8.G.7

8. Christine is fishing in the Gulf of Mexico. She rested her fishing pole on a stick in the ground. The tip of the fishing pole is 6 ft above the surface of the water. It is directly above the point where the land meets the edge of the water. There is 10 ft of fishing line between the tip of the pole and the surface of the water. How far from the edge of the water does the fishing line enter the water?

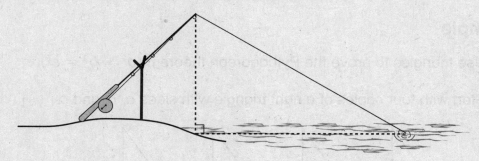

9. Tom built a ramp so that he could drive his motorcycle into the bed of his truck in order to take it to a motorcycle race. The tailgate of his truck is 3 ft from the ground. He made the ramp 5 ft long so it will fit in the bed of the truck. What is the distance from the bottom of the ramp to the point on the ground directly under the edge of the tailgate?

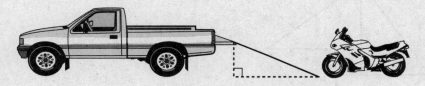

Explain how you were able to find the distance from the bottom of the ramp to the point on the ground directly under the edge of the tailgate.

_____

_____

_____

CCSS: 8.G.6

Lesson 26: Proofs of the Pythagorean Theorem

It is important to know how to prove the Pythagorean theorem.

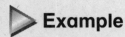 **Example**

Use triangles to prove the Pythagorean theorem ($a^2 + b^2 = c^2$).

Start with four copies of a right triangle with sides *a*, *b*, and *c*.

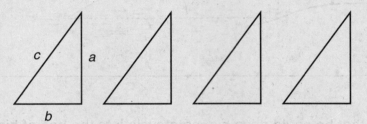

Arrange the triangles to make a square of side length *c*. At the center is a small square with side length $a - b$.

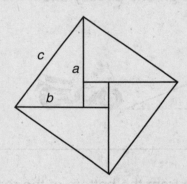

The area of the large square is c^2. This is equal to the sum of the areas of the four triangles and the small square.

area of four triangles	area of small square
$(\frac{1}{2}\,ab) \cdot 4 = 2ab$	$(a - b)^2 = a^2 - 2ab + b^2$

area of four triangles + area of small square = area of large square

$$2ab + a^2 - 2ab + b^2 = c^2$$

$$2\cancel{ab} + a^2 - 2\cancel{ab} + b^2 = c^2 \quad \text{Combine like terms.}$$

$$a^2 + b^2 = c^2$$

CCSS: 8.G.6

Proof of the Converse of the Pythagorean Theorem

The Pythagorean theorem states that if a triangle is a right triangle, then the sides of the triangle satisfy $a^2 + b^2 = c^2$. The converse of the Pythagorean theorem states that if the sides of a triangle satisfy $a^2 + b^2 = c^2$, then the triangle must be a right triangle. You can prove the converse of the Pythagorean theorem.

▶ **Example**

Triangle *ABC* has lengths of 18, 24, and 30, as shown below. Prove that the triangle must be a right triangle.

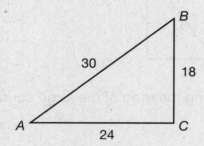

It is given that $AC^2 + BC^2 = AB^2$. You need to prove that $\triangle ABC$ is a right triangle.

Create right triangle *DEF* where *DF* = 24 and *EF* = 18.

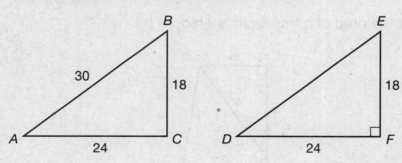

Because $\triangle DEF$ is a right triangle, $EF^2 + DF^2 = DE^2$. Substituting the lengths of *DF* and *EF*:

$EF^2 + DF^2 = DE^2$

$(18)^2 + (24)^2 = DE^2$

$324 + 576 = DE^2$

$900 = DE^2$

$30 = DE$

Because $AB = DE$ and $AC = DF$ and $BC = EF$, the triangles *ABC* and *DEF* have the same side lengths. That means they also have the same angle measures and $\angle ACB \cong \angle DFE$. Because $m\angle DFE = 90°$, $m\angle ACD = 90°$, which means triangle *ABC* is a right triangle.

⬤ Practice

Directions: Answer each question on the lines below.

1. Four congruent triangles can be arranged to make a square.

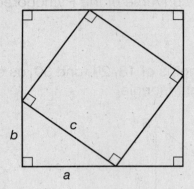

   Prove the Pythagorean theorem by calculating the area of the large square in two ways. Show your work.

2. The formula for the area of a trapezoid is $\frac{1}{2}h(b_1 + b_2)$.

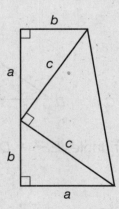

   Prove the Pythagorean theorem by calculating the area of the trapezoid in two ways. Show your work.

3. Prove that $\triangle DEF$ is a right triangle using the converse of the Pythagorean theorem.

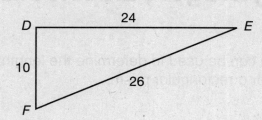

4. Prove that $\triangle LMN$ is a right triangle using the converse of the Pythagorean theorem.

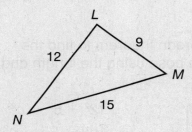

Lesson 27: Pythagorean Theorem in Three Dimensions

The Pythagorean theorem can be used to determine the length of a diagonal in a right prism, such as a cylinder or a rectangular prism.

Diagonals of Rectangular Prisms

To find the diagonal length of a rectangular prism, use the Pythagorean theorem to calculate the diagonal length of the base using the length and width of the base. Then you can use the Pythagorean theorem with that diagonal length and the height of the prism.

 Example

Find the length of the diagonal *D* of the following rectangular prism.

First use the Pythagorean theorem to find the diagonal length of the base using the length and width of the base.

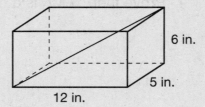

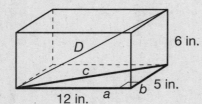

$$12^2 + 5^2 = c^2$$
$$144 + 25 = c^2$$
$$169 = c^2$$
$$13 = c$$

Now that you know the diagonal of the base is 13 in., you can use the Pythagorean theorem to find the diagonal length of the prism, *D*, using the diagonal of the base and the height of the prism.

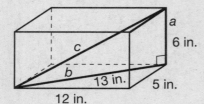

$$6^2 + 13^2 = c^2$$
$$36 + 169 = c^2$$
$$205 = c^2$$
$$\sqrt{205} = c$$

The length of the diagonal, *D*, is $\sqrt{205}$ in.

CCSS: 8.G.7

Diagonals of Cylinders

To find the length of the diagonal of a cylinder, use a diameter of the base and the height as two legs of a right triangle. The hypotenuse of the right triangle will be the diagonal of the cylinder.

 Example

Find the diagonal of the following cylinder.

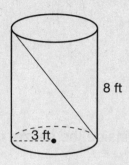

8 ft

3 ft

The radius of the cylinder is 3 ft. A diameter is twice the length of a radius. The diameter of the cylinder is therefore 6 ft. The height of the cylinder is 8 ft. The hypotenuse is the diagonal of the cylinder.

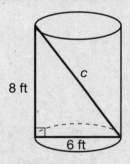

8 ft

c

6 ft

$6^2 + 8^2 = c^2$

$36 + 64 = c^2$

$100 = c^2$

$10 = c$

The length of the diagonal of the cylinder is 10 ft.

TIP: You can also use the following formula to find the diagonal length of a rectangular prism: $d = \sqrt{l^2 + w^2 + h^2}$, where *d* is the length of the diagonal, *l* is the length of the prism, *w* is the width of the prism, and *h* is the height of the prism. This formula is an extension of the Pythagorean theorem in three dimensions.

▷ **Example**

A cylindrical can of seltzer has a height of 5 inches and a radius of 1 inch, as shown.

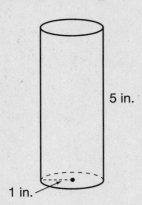

5 in.

1 in.

What is the longest straw that can fit completely inside the seltzer can?

To find the longest straw that will fit inside the can, you need to find the diagonal length of the cylindrical can. The diameter is twice the length of the radius. Therefore, the diameter is 2 in. The height is 5 in. The hypotenuse is the diagonal, which represents the maximum length of the straw in this example.

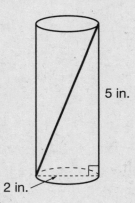

5 in.

2 in.

Once you know the lengths of the legs of a right triangle, use the Pythagorean theorem to find the hypotenuse.

$$2^2 + 5^2 = c^2$$
$$4 + 25 = c^2$$
$$29 = c^2$$
$$\sqrt{29} = c$$

The diagonal length is $\sqrt{29}$ in. Therefore, the longest straw that will fit inside the can will be $\sqrt{29}$ in.

⬤ Practice

Directions: For questions 1 through 4, find the diagonal length of each solid figure.

1.
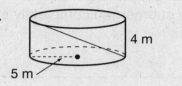
5 mm
8 mm   6 mm

_____

2.
4 m
5 m

_____

3.
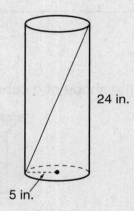
12 yd
4 yd   3 yd

_____

4.
24 in.
5 in.

_____

5. A juice box manufacturer wants to design a straw that cannot accidentally get stuck in the box.

15 cm

3 cm   5 cm

.If the length of the straw is an integer value, what is the smallest possible length of the straw?

_____

6. An oil tank is in the shape of a cylinder. A dipstick can be used to measure the amount of oil in the tank.

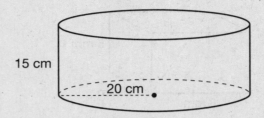

15 cm

20 cm

The dipstick has a length that is an integer value. What is the smallest possible length of a dipstick that cannot be submerged completely in the oil tank?

_____

7. A package is in the shape of a cube. The height of the package is 10 inches.

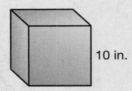

10 in.

What is the diagonal length of the package?

_____

Explain how you found the diagonal length of the package.

_____

_____

_____

Lesson 28: Volume

Volume (V) is the amount of space a solid takes up. It is measured in cubic units.

Cylinders

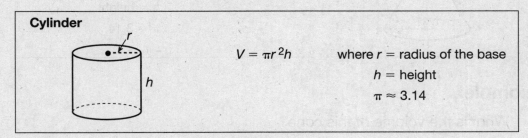

$$V = \pi r^2 h$$

where r = radius of the base

h = height

$\pi \approx 3.14$

▶ **Example**

What is the volume of this cylinder?

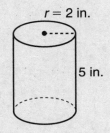

$r = 2$ in.

5 in.

Use the following formula.

$V = \pi r^2 h$

$\quad = 3.14 \cdot (2)^2 \cdot (5)$

$\quad = 3.14 \cdot 4 \cdot 5$

$\quad = 62.8$

The volume of the cylinder is approximately 62.8 in.^{3}.

CCSS: 8.G.9

Cones

To find the volume of a cone, use the following formula.

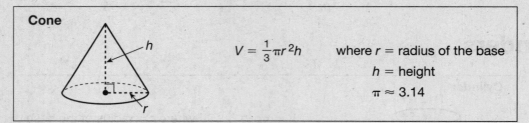

Cone

$V = \frac{1}{3}\pi r^2 h$ where r = radius of the base

h = height

$\pi \approx 3.14$

▷ Example

What is the volume of this cone?

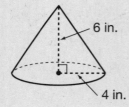

6 in.

4 in.

Use the following formula.

$V = \frac{1}{3}\pi r^2 h$

$= \frac{1}{3} \cdot 3.14 \cdot (4)^2 \cdot (6)$

$= \frac{1}{3} \cdot 3.14 \cdot 16 \cdot 6$

$= \frac{1}{3} \cdot 3.14 \cdot 96$

$= 100.48$

The volume of the cone is approximately 100.48 in.^{3}.

Spheres

To find the volume of a sphere, use the following formula.

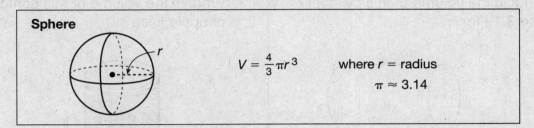

Sphere

$$V = \frac{4}{3}\pi r^3 \qquad \text{where } r = \text{radius}$$
$$\pi \approx 3.14$$

▷ Example

What is the volume of the sphere below?

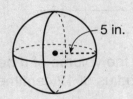

5 in.

Use the following formula.

$$V = \frac{4}{3}\pi r^3$$

$$= \frac{4}{3} \cdot 3.14 \cdot (5)^3$$

$$= \frac{4}{3} \cdot 3.14 \cdot 125$$

$$= 523.\overline{3}$$

The volume of the sphere is approximately 523.3 in.[3].

Practice

Directions: Answer each question.

1. What is the volume of this cylinder? Use 3.14 for π.

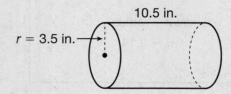

r = 3.5 in.

10.5 in.

A. 192.325 in.^{3}

B. 307.72 in.^{3}

C. 403.8825 in.^{3}

D. 2307.9 in.^{3}

2. How much ice cream can fit inside a cone that has a diameter of 8 centimeters and a height of 9 centimeters?

A. 48 cm^3

B. 150.72 cm^3

C. 192 cm^3

D. 602.88 cm^3

3. What is the volume of this container of apple juice?

3 in.

APPLE JUICE

7 in.

$V = $ _____

4. What is the volume of a beach ball with a radius of 12 centimeters?

$V = $ _____

5. The smallest object in space that is spherical due to its own gravity is Mimas, one of the moons of Saturn. The radius of Mimas is approximately 200 km. What is the approximate volume of the moon, to the nearest million cubic km?

$V = $ _____

166

6. A party hat is in the shape of a cone with a radius of 3 in. and a height of 5 in. What is the volume of the party hat? Use 3.14 for π.

V = _____

7. What is the volume of the following cylinder? Use 3.14 for π and round your answer to the nearest whole number.

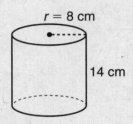

r = 8 cm

14 cm

V = _____

8. A sphere has a diameter of 2 m. What is the volume of the sphere? Use 3.14 for π and round your answer to the nearest tenth.

V = _____

9. What is the volume of the following cone?

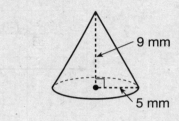

9 mm

5 mm

V = _____

10. A cylindrical gas tank has a capacity of approximately 157 m³. Its radius is 5 m.

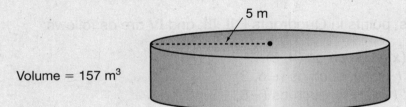

5 m

Volume = 157 m³

What is the height of the cylindrical gas tank? _____

Explain how you found the height of the gas tank.

_____

_____

_____

CCSS: 8.G.3, 8.G.8

Lesson 29: Coordinate Geometry

The coordinate plane has a horizontal axis (called the **x-axis**) and a vertical axis (called the **y-axis**). The point where the two axes intersect is called the **origin**. The coordinate plane is divided into four quadrants.

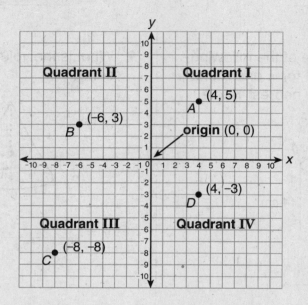

Points on the coordinate plane are described by ordered pairs (also known as coordinates). The first number of an ordered pair describes a point's location on the x-axis. The second number describes its location on the y-axis.

In general terms, points in Quadrants I, II, III, and IV are as follows:

Quadrant I: (x, y), such as (4, 5)
Quadrant II: (−x, y), such as (−6, 3)
Quadrant III: (−x, −y), such as (−8, −8)
Quadrant IV: (x, −y), such as (4, −3)

CCSS: 8.G.3, 8.G.8

Using the Pythagorean Theorem with Coordinate Geometry

You can use the Pythagorean theorem to find the distance between any two points on a coordinate plane.

▷ **Example**

What is the distance between points A and B? Round to the nearest tenth.

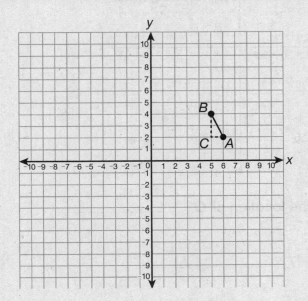

Form a right triangle where $\overline{AB}$ is the hypotenuse. Label this new point C. The legs of the right triangle are $\overline{AC}$ and $\overline{BC}$.

Use the x-coordinate of points A and C to find the distance between points A and C. The length of $\overline{AC}$ is $|6 - 5| = 1$ unit.

Use the y-coordinate of points B and C to find the distance between points B and C. The length of $\overline{BC}$ is $|4 - 2| = 2$ units.

Substitute the values into the Pythagorean theorem.

$$AB^2 = AC^2 + BC^2$$
$$AB^2 = (1)^2 + (2)^2$$
$$AB^2 = 1 + 4$$
$$AB = \sqrt{5} \qquad \text{Take the square root of both sides.}$$
$$AB \approx 2.3$$

The distance between points A and B is about 2.3 units.

169

◉ **Practice**

Directions: Use the following coordinate plane to answer questions 1 through 8.

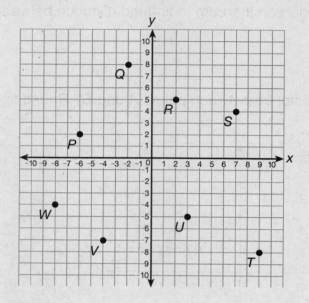

1. What ordered pair describes the location of *Q*? _____

2. What point is located at (−4, −7)? _____

3. In what quadrant is *P* located? _____

4. What point is located at (3, −5)? _____

5. What ordered pair describes the location of *S*? _____

6. Name two points that are located in Quadrant III. _____

7. What ordered pair describes the location of *T*? _____

8. What point is located at (2, 5)? _____

9. What is the distance between points *E* and *F*? Round to the nearest tenth.

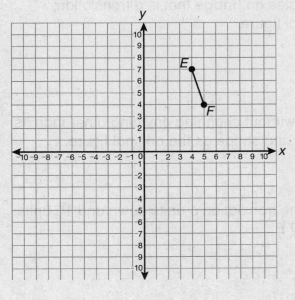

EF ≈ _____

10. What is the distance between points *M* and *N*? Round to the nearest tenth.

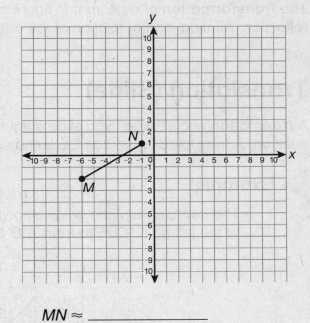

MN ≈ _____

11. Given points *C*(−6, 10) and *D*(−3, −2), what is the length of $\overline{CD}$? Round to the nearest tenth.

CD ≈ _____

12. Given points *S*(−4, −2) and *T*(−1, 0), what is the length of $\overline{ST}$? Round to the nearest tenth.

ST ≈ _____

Explain how you identified the length of $\overline{ST}$ to the nearest tenth.

_____

_____

_____

Lesson 30: Transformations

The **transformation** of a geometric figure creates an **image** that is a translation, reflection, rotation, and/or dilation of the original figure.

Translation (Slide)

A **translation** occurs when you slide a figure, without changing anything other than its position. A translated figure has the same size and shape as the original figure.

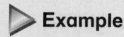 **Example**

The coordinate plane below shows the translation 6 units down and 10 units to the left of *ABCDE* to form *A′B′C′D′E′*.

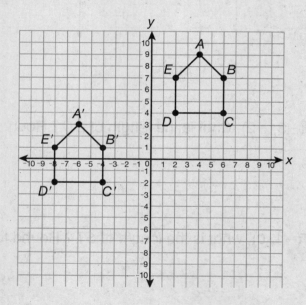

To form *A′B′C′D′E′*, the *x*-coordinate of each vertex of *ABCDE* was decreased by 10. The *y*-coordinate of each vertex of *ABCDE* was decreased by 6.

Vertices of *ABCDE*
A: (4, 9)
B: (6, 7)
C: (6, 4)
D: (2, 4)
E: (2, 7)

Vertices of *A′B′C′D′E′*
A′: (−6, 3)
B′: (−4, 1)
C′: (−4, −2)
D′: (−8, −2)
E′: (−8, 1)

CCSS: 8.G.3

Reflection (Flip)

A **reflection** occurs when you flip a figure over a given line and its mirror image in that line is created. A reflected figure has the same size and shape as the original figure, but a different orientation.

▷ **Example**

The following coordinate plane shows the reflection over the *y*-axis of trapezoid *MNOP* to form trapezoid *M′N′O′P′*.

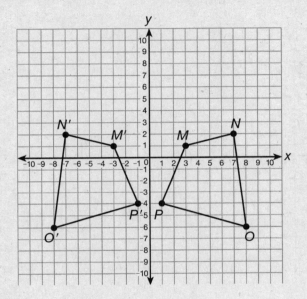

To form trapezoid *M′N′O′P′*, the *y*-coordinate of each vertex of trapezoid *MNOP* stayed the same. The *x*-coordinate of each vertex of trapezoid *MNOP* became its opposite in *M′N′O′P′*.

Vertices of trapezoid *MNOP*
M: (3, 1)
N: (7, 2)
O: (8, −6)
P: (1, −4)

Vertices of trapezoid *M′N′O′P′*
M′: (−3, 1)
N′: (−7, 2)
O′: (−8, −6)
P′: (−1, −4)

Rotation (Turn)

A **rotation** occurs when you turn a figure around a given point. Figures can be rotated in a clockwise or counterclockwise direction. A rotated figure has the same size and shape as the original figure, but a different orientation.

▶ **Example**

The following coordinate plane shows the 90°-clockwise rotation around *J* of △*JKL* to form △*JK'L'*.

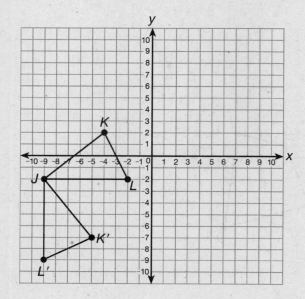

Vertices of △*JKL*	Vertices of △*JK'L'*
J: (−9, −2)	*J*: (−9, −2)
K: (−4, 2)	*K'*: (−5, −7)
L: (−2, −2)	*L'*: (−9, −9)

 TIP: A rotated figure may be rotated around the origin or a point in the figure. If the figure is rotated around a point in the figure (as shown in the example above), the coordinates of that point will not change.

CCSS: 8.G.3

Dilation (Scale)

A **dilation** occurs when you enlarge or reduce a figure from the origin. To perform a dilation on a figure, multiply the coordinates of each vertex by a positive **scale factor**, k. If the scale factor is less than 1 ($k < 1$), the dilation will be a **reduction**. If the scale factor is greater than 1 ($k > 1$), the dilation will be an **enlargement**.

A dilated figure has the same shape as the original figure. The area of a dilated figure will be k^2 times the area of the original figure.

 Example

The following coordinate plane shows a dilation, using a scale factor of $\frac{2}{3}$, of rectangle $WXYZ$ to form rectangle $W'X'Y'Z'$.

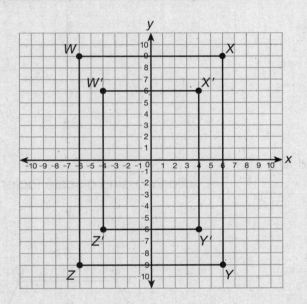

To form rectangle $W'X'Y'Z'$, the x- and y-coordinates of each vertex of rectangle $WXYZ$ have been multiplied by the scale factor of $\frac{2}{3}$. Rectangle $W'X'Y'Z'$ is a reduction of rectangle $WXYZ$.

Vertices of rectangle $WXYZ$
W: $(-6, 9)$
X: $(6, 9)$
Y: $(6, -9)$
Z: $(-6, -9)$

Vertices of rectangle $W'X'Y'Z'$
W': $(-4, 6)$
X': $(4, 6)$
Y': $(4, -6)$
Z': $(-4, -6)$

The area of $WXYZ$ is 216 square units. The area of $W'X'Y'Z'$ is $\left(\frac{2}{3}\right)^2 \cdot 216 = 96$ square units.

175

CCSS: 8.G.3

 Practice

Directions: For questions 1 through 8, draw the given transformation of each figure.

1. reflection over the *y*-axis

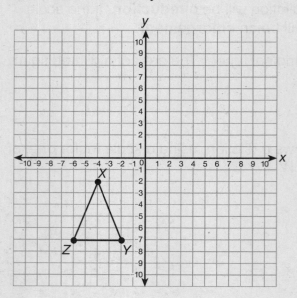

2. 180°-clockwise rotation around *M*

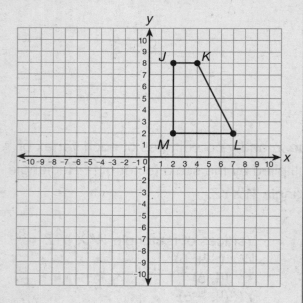

3. translation 8 units left and 7 units down

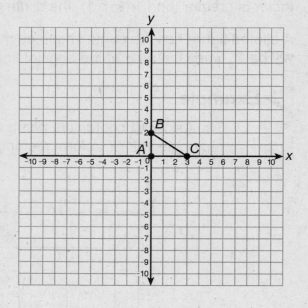

4. dilation using a scale factor of $\frac{1}{2}$

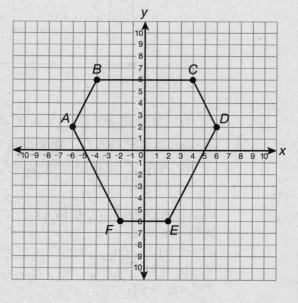

CCSS: 8.G.3

5. translation 10 units right

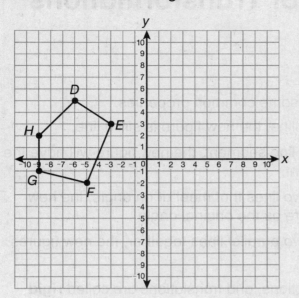

7. reflection over the *x*-axis

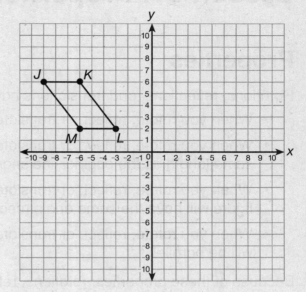

6. 90°-counterclockwise rotation around *R*

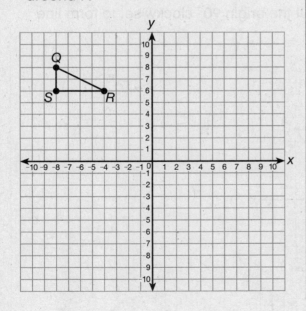

8. dilation using a scale factor of 2

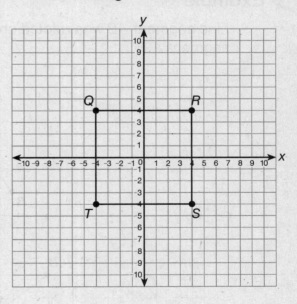

9. Explain how you find the new coordinates for the vertices of a dilated figure.

_____

_____

_____

CCSS: 8.G.1a, 8.G.1b, 8.G.1c

Lesson 31: Properties of Transformations

Properties

Rotations, reflections, and translations have some common properties.

- When you rotate, reflect, or translate a line, the new figure is still a line.

- When you rotate, reflect, or translate a line segment, the new figure will have the same length as the original line segment.

- When you rotate, reflect, or translate two lines that meet at an angle, the new figure will have the same angle measure as the original angle.

- When you rotate, reflect, or translate two parallel lines together, the new figures will be parallel.

Because of these properties, rotations, reflections, and translations are called **rigid motions**. They take the original figure and rigidly move it to a new location.

 Example

Line segment *AB* was rotated about the origin 90° clockwise, to form line segment *A′B′*.

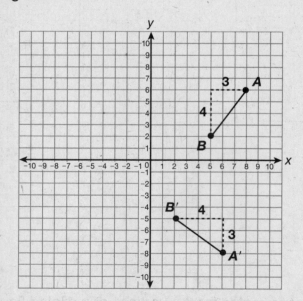

First note that line segment *A′B′* is in fact a line. Both segments have the same length.

$$AB^2 = (4)^2 + (3)^2$$
$$AB^2 = 16 + 9$$
$$AB^2 = 25$$
$$AB = 5$$

$$A'B'^2 = (4)^2 + (3)^2$$
$$A'B'^2 = 16 + 9$$
$$A'B'^2 = 25$$
$$A'B' = 5$$

CCSS: 8.G.1a, 8.G.1b, 8.G.1c

▷ **Example**

Parallel lines *m* and *n* were reflected over the *y*-axis, and then translated 1 unit down and 4 units to the right, to form lines *m'* and *n'*.

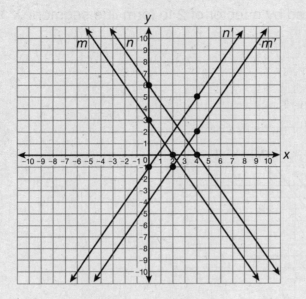

Lines *m* and *n* have the same slope and are parallel.

Slope of $m = \dfrac{3-0}{0-2} = -\dfrac{3}{2}$

Slope of $n = \dfrac{6-0}{0-4} = -\dfrac{3}{2}$

You can confirm that lines *m'* and *n'* are also parallel.

Slope of $m' = \dfrac{2-(-1)}{4-2} = \dfrac{3}{2}$

Slope of $n' = \dfrac{5-(-1)}{4-0} = \dfrac{3}{2}$

179

CCSS: 8.G.1a, 8.G.1b, 8.G.1c

Dilations have similar properties, but line segments that are dilated do not have the same length.

▷ Example

Line segment *XY* is dilated by a factor of 2 to form line segment *X′ Y′*.

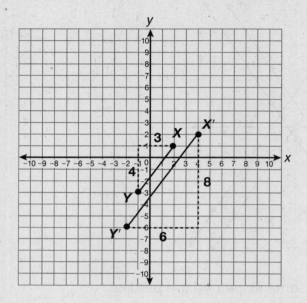

The line segments have different lengths.

$XY^2 = (4)^2 + (3)^2$ $X′ Y′^2 = (8)^2 + (6)^2$

$XY^2 = 16 + 9$ $X′ Y′^2 = 64 + 36$

$XY^2 = 25$ $X′ Y′^2 = 100$

$XY = 5$ $X′ Y′ = 10$

CCSS: 8.G.1a, 8.G.1b, 8.G.1c

Practice

Directions: Perform the transformation of the figure, and find the measurement.

1. Reflect over the *y*-axis.

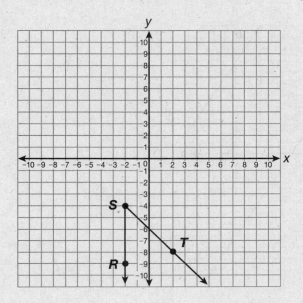

Measure of angle *R′S′T′*

_____

3. Rotate 90° counterclockwise around the origin.

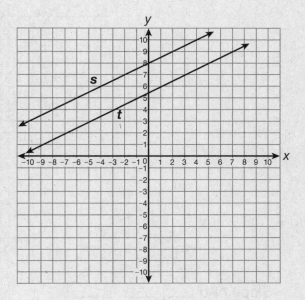

Slope of *s′* _____

Slope of *t′* _____

2. Translate 2 units down, 3 units right.

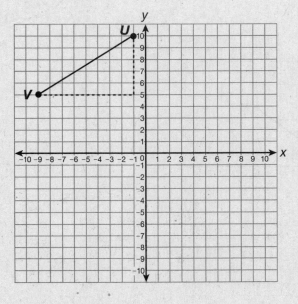

Length of *U′V′* _____

4. Reflect over the *x*-axis.

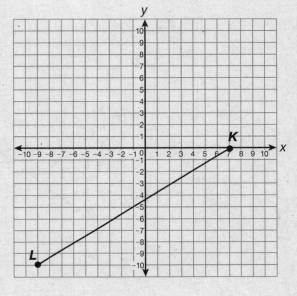

Length of *K′L′* _____

181

CCSS: 8.G.1a, 8.G.1b, 8.G.1c

Directions: Perform the transformations of the figure, and find the desired measurement.

5. Rotate about the origin 90° clockwise, and reflect over the *y*-axis.

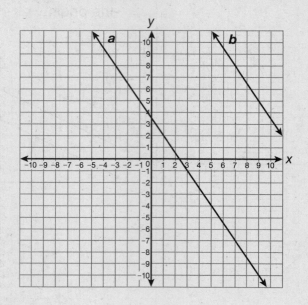

Slope of *a'* _____

Slope of *b'* _____

6. Reflect over the *y*-axis, reflect over the *x*-axis, and rotate 180° clockwise about the origin.

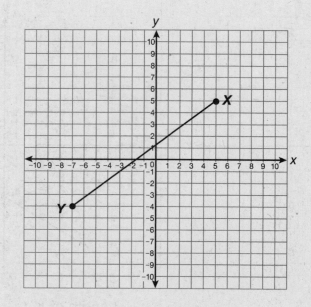

Length of *X' Y'* _____

182

CCSS: 8.G.2

Lesson 32: Congruent Figures

Two figures are **congruent** if they have the same size and shape. Congruent figures have corresponding sides with the same length, and corresponding angles with the same measure. If two figures are congruent, you can take one figure, change it with a sequence of rigid motions, and get the other figure. It doesn't matter which figure you start with; either figure will transform to the other figure.

▶ **Example**

Find the sequence of transformations that take △*EFG* to △*HIJ*.

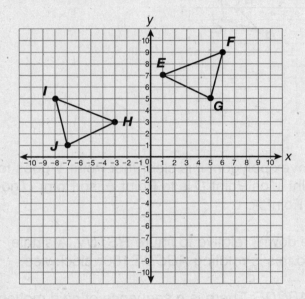

You can identify which sides correspond by identifying the slope of each side:

Slope of $\overline{EF} = \frac{2}{5}$ Slope of $\overline{HI} = -\frac{2}{5}$

Slope of $\overline{EG} = -\frac{1}{2}$ Slope of $\overline{HJ} = \frac{1}{2}$

Slope of $\overline{FG} = 4$ Slope of $\overline{IJ} = -4$

You can see that the slopes of the line segments of one figure are the opposites of the slopes of the line segments of the other figure. You can use this information as a clue to make your transformations.

Start by reflecting △*EFG* along the *y*-axis. Then take the new figure and translate it 4 units down and 2 units to the left.

△*EFG* is congruent to △*HIJ*. There is a sequence of rigid motions that take △*EFG* to △*HIJ*.

Example

Find the sequence of transformations that take *RSTUV* to *ABCDE*.

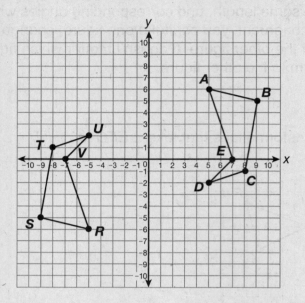

Because *RSTUV* is a convex polygon, it is clear that Point *V* goes to Point *E*. Point *U* is as close to Point *V* as Point *D* is to Point *E*, so Point *U* goes to Point *D*. Point *R* goes to Point *A*.

Figure *ABCDE* is "upside down" relative to figure *RSTUV*. A single transformation that will do this is rotating *ABCDE* 180° clockwise.

RSTUV is congruent to *ABCDE*. There is a sequence of rigid motions that take *RSTUV* to *ABCDE*.

TIP: Another way to transform *RSTUV* to *ABCDE* is to reflect *RSTUV* across the *x*- and *y*-axis.

CCSS: 8.G.2

 Practice

Directions: For questions 1 through 4, identify the series of transformations that take the given figure to the other figure.

1. Given figure *WXYZ*.

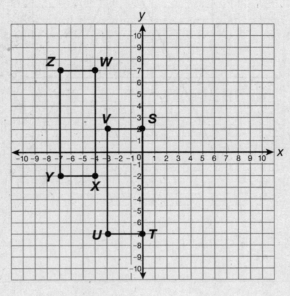

_____

_____

2. Given △*DEF*.

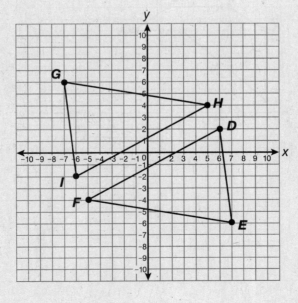

_____

_____

185

3. Given △*GHI*.

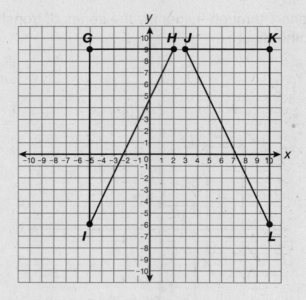

_____

_____

4. Given figure *MNOPQR*.

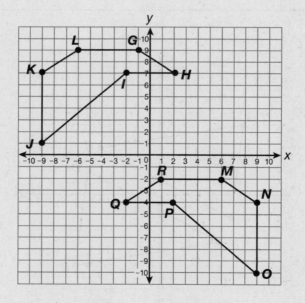

_____

_____

CCSS: 8.G.4

Lesson 33: Similar Figures

Two figures are **similar** if they have the same shape. Similar figures have corresponding angles with the same measures. Similar figures may have corresponding sides with different lengths, but corresponding sides always have the same ratio.

Any sequence of rotations, reflections, translations, and dilations will not affect the similarity of a figure and its image. That is because rotations, reflections, translations, and dilations do not affect the shape of a figure.

 Example

ABCD has vertices at A (−10, 10), B (−4, 10), C (−4, 6), and D (−10, 6), and FGHI has vertices at F (2, 5), G (5, 5), H (5, 3), and I (2, 3). Determine whether ABCD is similar to FGHI by identifying a series of transformations that takes one figure to another.

ABCD can be reflected over the y-axis and then dilated by a scale factor of $\frac{1}{2}$. The resulting figure will be FGHI.

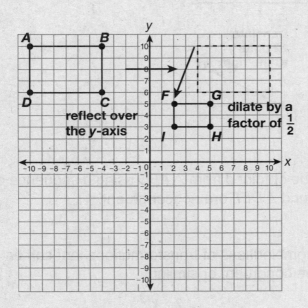

Because FGHI can be obtained from ABCD using a sequence of transformations, ABCD and FGHI are similar.

 TIP: Congruent figures are also similar figures because they have the same shape. However, similar figures are not congruent unless they are the same size.

You can also use a sequence of transformations to determine that two figures are not similar.

 Example

Triangle *ABC* has vertices at *A*(−2, −4), *B*(−2, −6), and *C*(−6, −6), and △*DEF* has vertices at *D*(9, −6), *E*(9, −9), and *F*(3, −9). Determine whether △*ABC* is congruent to △*DEF*.

Triangle *ABC* can be enlarged with a dilation of $\frac{3}{2}$ and then translated 12 units to the right. The resulting figure will match △*DEF*.

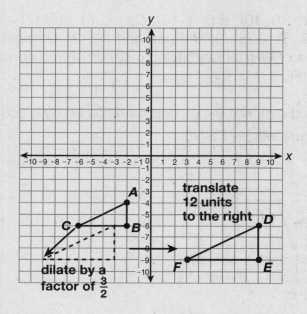

Because △*DEF* cannot be obtained from △*ABC* using a sequence of rigid transformations, △*ABC* and △*DEF* are not congruent.

 TIP: There may be several combinations of transformations that can be used to determine whether figures are congruent or similar.

CCSS: 8.G.4

Practice

Directions: For questions 1 through 4, write a sequence of transformations that can be used to show that the two given figures are similar.

1.

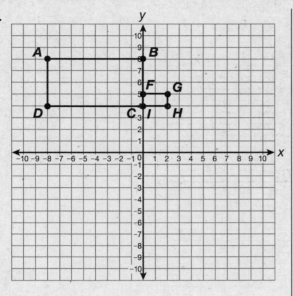

ABCD to *FGHI* _____

3.

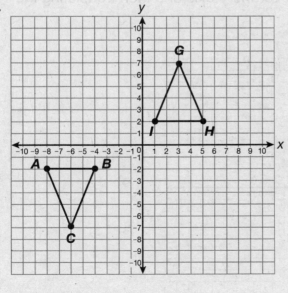

ABCDEF to *UVWXYZ* _____

2.

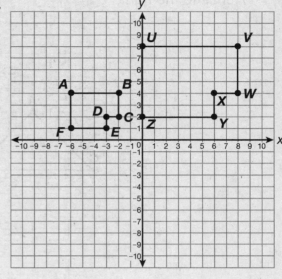

ABC to *GHI* _____

4.

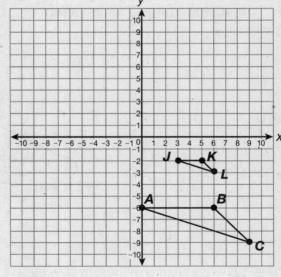

ABC to *JKL* _____

189

Directions: For questions 5 and 6, draw a similar figure based on the given transformations.

5. Draw similar △*DEF* by dilating △*ABC* by $\frac{1}{5}$ and then translating the resulting image 7 units to the right.

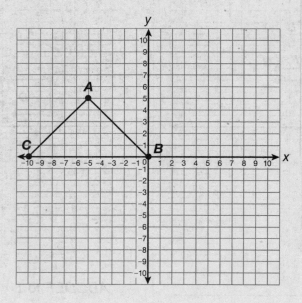

6. Draw similar △*QRS* by reflecting △*GHI* over the *y*-axis and then translating the resulting image 7 units down.

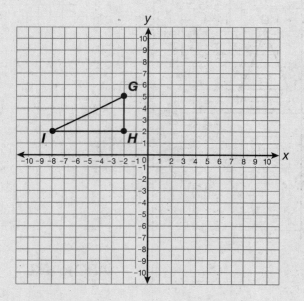

Lesson 34: Angles

Parallel Lines

When two parallel lines are cut by a transversal, certain angle relationships are formed. In the following diagram, parallel lines *m* and *n* have been cut by transversal *t*.

Given: $m \parallel n$

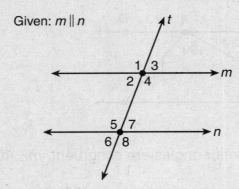

Vertical angles are congruent ($\cong$) angles that are formed by two intersecting lines (congruent angles have the same measure).

$\angle 1 \cong \angle 4 \qquad \angle 2 \cong \angle 3 \qquad \angle 5 \cong \angle 8 \qquad \angle 6 \cong \angle 7$

Interior angles lie between two parallel lines that are cut by a transversal.
$\angle 2$, $\angle 4$, $\angle 5$, and $\angle 7$ are interior angles.

Exterior angles lie outside two parallel lines that are cut by a transversal.
$\angle 1$, $\angle 3$, $\angle 6$, and $\angle 8$ are exterior angles.

Alternate interior angles are interior angles that lie on opposite sides of a transversal. If the lines are parallel, then the alternate interior angles are congruent.
$\angle 2 \cong \angle 7 \qquad \angle 4 \cong \angle 5$

Alternate exterior angles are exterior angles that lie on opposite sides of a transversal. If the lines are parallel, then the alternate exterior angles are congruent.
$\angle 1 \cong \angle 8 \qquad \angle 3 \cong \angle 6$

Corresponding angles lie on the same side of a transversal, in corresponding positions with respect to the two lines that the transversal intersects. If the lines are parallel, then the corresponding angles are congruent.
$\angle 1 \cong \angle 5 \qquad \angle 3 \cong \angle 7 \qquad \angle 2 \cong \angle 6 \qquad \angle 4 \cong \angle 8$

Angle Sum of Triangles

The angle sum of a triangle is always 180°. One way to show this is to make a line that goes through one vertex of the triangle and is parallel to the opposite side.

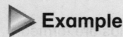 **Example**

Lines *AD* and *BC* are parallel. What are the measures of ∠*ABC* and ∠*ACB*?

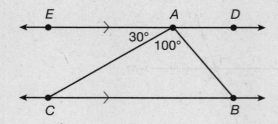

Since alternate interior angles are congruent, *m*∠*ACB* = 30°.

The angle measure of a straight line is 180°. Using this information, find *m*∠*DAB*.

$$180° = m\angle DAB + m\angle BAC + m\angle CAE$$
$$180° = m\angle DAB + (100°) + (30°)$$
$$180° = m\angle DAB + 130°$$
$$\ \ 50° = m\angle DAB$$

Since alternate interior angles are congruent, *m*∠*ABC* = 50°.

Exterior Angles of Triangles

Exterior angles are the angles between any side of a triangle and a line extended from another side. An adjacent interior angle and its exterior angle lie on a straight line, so their angle sum is 180°. The exterior angle is equal to the sum of the other two interior angles. You can also use parallel lines to find the exterior angle of a triangle.

▶ **Example**

Lines *FG* and *HJ* are parallel. What is the measure of exterior ∠*FHI*?

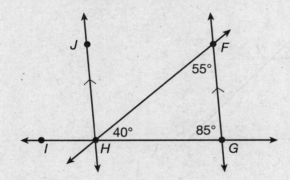

∠*FHI* can be broken up into two angles, ∠*FHJ* and ∠*JHI*. To find *m*∠*FHI*, find the measures of ∠*FHJ* and ∠*JHI*.

Since alternate interior angles are congruent, *m*∠*FHJ* = 55°.

Corresponding angles are congruent, so *m*∠*JHI* = 85°.

$$m\angle FHI = m\angle FHJ + m\angle JHI$$
$$= 55° + 85°$$
$$= 140°$$

CCSS: 8.G.5

Angle-Angle Similarity

The **angle-angle similarity postulate** states that if two pairs of angles in two triangles are congruent, then the triangles are similar. Given a triangle you can create a similar triangle by drawing a line parallel to a side, inside the triangle.

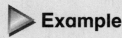 **Example**

Line *QN* is parallel to line *PO*. Is △*MNQ* similar to △*MOP*?

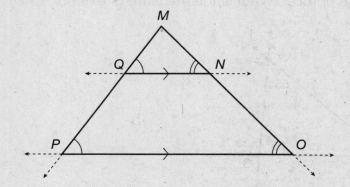

Since corresponding angles are congruent, ∠*MNQ* ≅ ∠*MOP* and ∠*MQN* ≅ ∠*MPO*. Since ∠*M* is common to both triangles, it is congruent to itself.

Because all the angles are congruent, △*MNQ* is similar to △*MOP*.

⬤ Practice

Directions: Use the following diagram to answer questions 1 through 8.

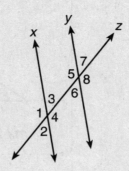

1. List all pairs of vertical angles. _____

2. List all pairs of corresponding angles. _____

3. List all pairs of alternate interior angles. _____

4. List all pairs of alternate exterior angles. _____

5. List two pairs of congruent angles. _____

6. List two pairs of supplementary angles. _____

7. If $m\angle 5 = 135°$, $m\angle 8 =$ _____

8. If $m\angle 2 = 45°$, $m\angle 4 =$ _____

Directions: For questions 9 through 12, find the measure of the given angle.

9.

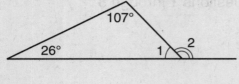

$m\angle 2 =$ _____

11.

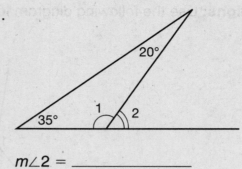

$m\angle 2 =$ _____

10.

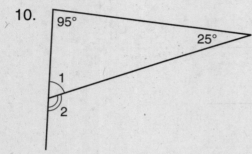

$m\angle 2 =$ _____

12.

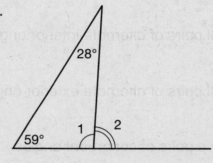

$m\angle 2 =$ _____

Directions: Use the following figure to answer questions 13 through 21.

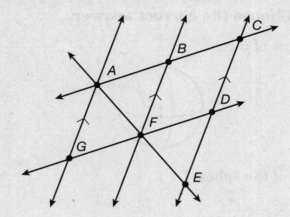

13. Is ∠*AGF* congruent to ∠*BFD*? _____

14. Is ∠*EDF* congruent to ∠*BFD*? _____

15. Is ∠*AFG* congruent to ∠*DFE*? _____

16. Is ∠*AFB* congruent to ∠*DEF*? _____

17. Is ∠*BCD* congruent to ∠*ABF*? _____

18. Is ∠*DEF* congruent to ∠*FBA*? _____

19. Is △*ABF* similar to △*DEF*? _____

20. Is △*ABF* similar to △*AFG*? _____

21. Is △*AFG* similar to △*DEF*? _____

Unit 4 Practice Test

Read each question. Choose the correct answer.

1. A sphere has a radius of 6 cm.

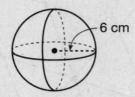

What is the volume of the sphere?

A. 216 cm^3

B. 288 cm^3

C. $216\pi \text{ cm}^3$

D. $288\pi \text{ cm}^3$

2. Line segment AB with points $A(-3, 6)$ and $B(9, 1)$ is shown on the coordinate grid below.

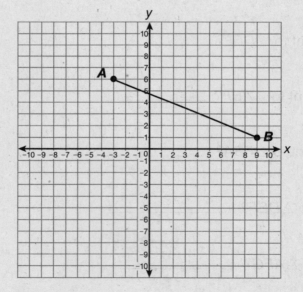

What is the length of $\overline{AB}$?

_____

3. A barrel is in the shape of a cylinder with a height of 4 ft and a radius of 1 ft.

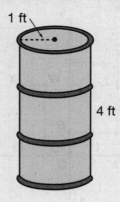

1 ft

4 ft

Using 3.14 for π, what is the volume of the barrel?

A. 4 ft³

B. 8.14 ft³

C. 12.56 ft³

D. 25.12 ft³

4. A box in the shape of a rectangular prism has dimensions 4 in. by 4 in. by 7 in. What is the length of the diagonal of the box?

_____

5. A soup can is in the shape of a cylinder. The radius of the soup can is 1.5 inches, and its height is 5 inches.

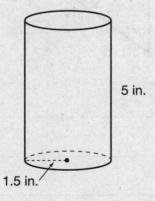

5 in.

1.5 in.

Using 3.14 for π, what is the volume of the soup can?

_____

6. A sphere has a radius of 12 cm. What is the volume of the sphere?

_____

7. Rectangles *ABCD* and *XYZW* are shown on the following coordinate grid.

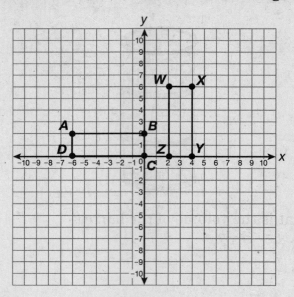

What transformations take *ABCD* to *XYZW*? _____

8. Prove that △*DEF* is a right triangle using the converse of the Pythagorean theorem.

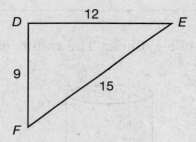

_____

_____

_____

9. A box is in the shape of a rectangular prism. The box has a length of 8 in., a height of 5 in., and a width of 4 in., as shown below.

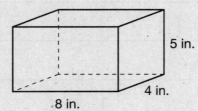

5 in.

4 in.

8 in.

A person wants to ship a dowel in the box. The dowel is a long, thin piece of wood. The length of the dowel is a whole number of inches. What is the greatest length that the dowel can be to fit inside the box?

_____

10. After a dilation, are two parallel lines still parallel? Explain.

_____

_____

_____

11. Ishmael rides in a hot air balloon that is tethered to the ground with a long rope. The length of the rope is 250 feet. The wind blows the balloon so that it forms a right triangle with the ground, as shown below. The balloon floats 200 feet away from the spot where it is tied to the ground.

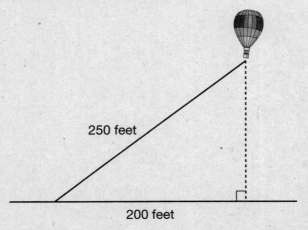

250 feet

200 feet

How high is the balloon floating above the ground?

_____

12. Why do dilations produce similar figures, but not congruent figures?

_____

_____

_____

13. Triangle *FGH* is plotted at *F*(−2, 7), *G*(1, 10), and *H*(3, 2).

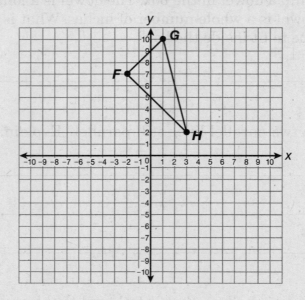

Triangle *FGH* is then reflected over the *x*-axis to form △*F′G′H′*. What are the coordinates of △*F′G′H′*?

F′: _____

G′: _____

H′: _____

14. Given that *m* is parallel to *n*, and $m\angle 1$ is 142°, what is $m\angle 2$?

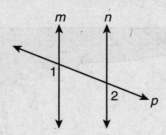

15. Name the transformations that take *ABCD* to *A'B'C'D'*.

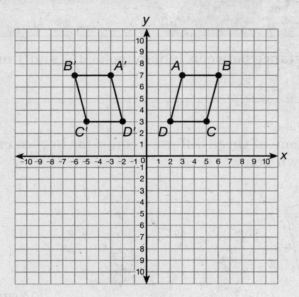

16. Columbus, OH, is about 830 miles north of Tampa, FL. Indianapolis, IN, is about 170 miles west of Columbus.

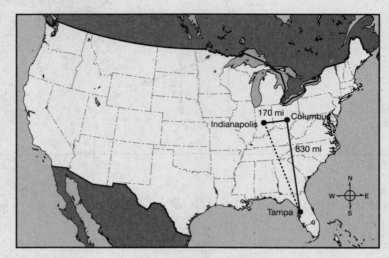

Approximate the distance between Indianapolis, IN, and Tampa, FL.

_____

17. Isabella can choose one of three different cones for her ice cream. The dimensions of the cones are shown below.

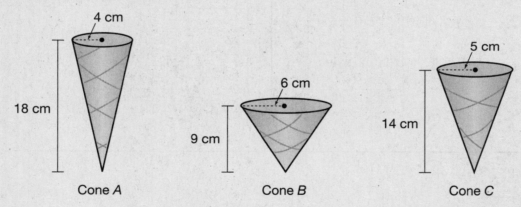

Isabella wants to choose a cone that holds the greatest amount of ice cream. Which cone should she choose?

_____

Explain your answer.

_____

_____

_____

18. The measure of two angles of △ABC are given as 40° and 80°. The measures of two angles of △FED are given as 80° and 60°.

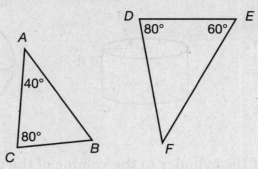

Part A

Are △ABC and △DEF similar? _____

Part B

Explain how you are able to tell whether or not △ABC and △DEF are similar.

_____

_____

_____

19. The following figures all have a radius of 1 foot. The heights of the cylinder and cone are also 1 foot.

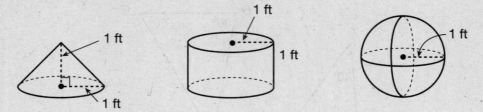

Part A
Compare the volume of the cylinder to the volume of the cone.

_____

Part B
Compare the volume of the cone to the volume of the sphere.

_____

Part C
Compare the volume of the sphere to the volume of the cylinder.

_____

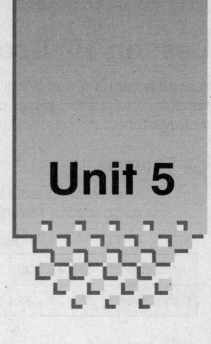

Unit 5

Statistics and Probability

Statistics are an invaluable tool when making decisions based on data. The Department of Transportation may base a decision about whether to repave a road on the mean number of vehicles that use the road on a weekly basis. Newspapers and magazines frequently use graphs to present some aspect of the news, such as how the government is spending tax dollars or how the season's top movies compare at the box office. When playing a board game, you may determine the odds of landing on a certain space.

In this unit, you will make and use scatter plots to identify patterns. You will interpret the slope and *y*-intercept of the model. Finally, you will make and use two-way tables to summarize data.

In This Unit
Scatter Plots

Trend Lines

Interpreting Linear Models

Two-Way Tables

Lesson 35: Scatter Plots

A **scatter plot** is a graph that plots the values of two variables as ordered pairs. You can then analyze the scatter plot by examining how closely the points come to forming a straight line.

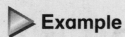 **Example**

The following table shows the number of passes Brett attempted and completed in 16 football games.

Pass Attempts Versus Completions

Attempts	28	31	39	35	34	35	27	32	29	26	42	27	39	28	29	30
Completions	22	20	25	20	27	21	16	19	16	18	24	15	21	18	18	15

The following scatter plot displays the data from the table.

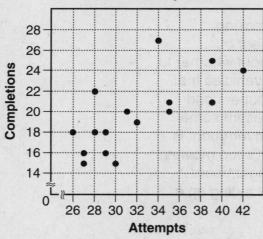

**Pass Attempts
Versus Completions**

From the scatter plot you can tell that the number of completions increases as the number of pass attempts increase.

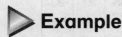

The closer the data points are to forming a straight line, the greater the correlation between the variables. The grouping of points along the line is called **clustering**. When the data points are not clustered closely together along this imaginary line, there is not a strong association between the data sets.

However, even scatter plots with closely clustered data points can have a few points that are far from the line. These points are called **outliers** because they lie outside the general pattern of the data. Plotting data points in a scatter plot helps identify outliers because the outliers will visibly stick out from the other data points.

▶ Example

The following table and scatter plot shows the number of hours 16 students studied for a test and the number of questions they got right in a 30-question test.

Time Studied Versus Questions Right

Questions Right	29	14	12	15	19	24	24	27	11	22	26	8	18	30	23	21
Hours Studied	10	2	1	2	3	4	5	7.5	0.5	8.5	2.5	0	2.5	9.5	5	4

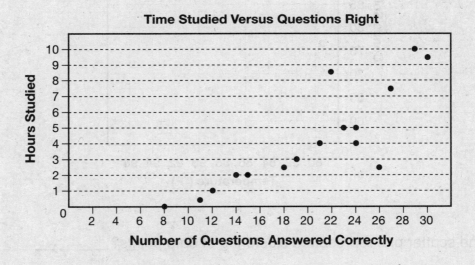

The data points are mostly clustered together in a line that extends from (8, 0) to (30, 9.5). However, there are two points that are far from this line: (22, 8.5) is very high above the line and (26, 2.5) is very low beneath the line on the scatter plot. These are the outliers of the data.

Practice

Directions: Use the following information to answer questions 1 through 4.

Christina works at the ice cream shop during summer vacation. She uses the following table to record the highest temperature each day for two weeks and the number of ice cream cones she sold on each of those days.

Temperature Versus Cones Sold

Temperature (°F)	85	87	91	95	88	83	80	82	88	90	93	85	87	83
Cones Sold	76	77	70	100	91	79	67	73	78	87	92	95	85	68

1. Use the information from the table to create a scatter plot of the data.

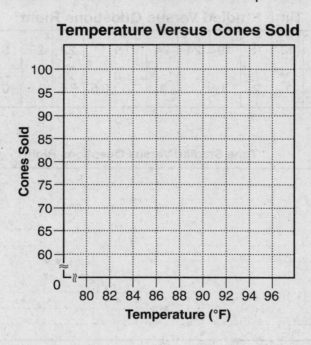

2. Does the scatter plot represent a clustering of data points? _____

3. Is there a relationship between the high temperature and the number of ice cream cones sold each day?

   _____

4. Are there any outliers in the data? If so, what are the ordered pairs of the point(s)?

CCSS: 8.SP.1

Directions: Use the following information to answer questions 5 through 8.

The following table shows the relationship between the average amounts of television 12 students watch per week and their grade point averages. The grade point average has a maximum value of 4.0.

Hours of TV Watched Versus Grade Point Average

Hours of TV Watched	9	6	19	26	12	4	28	21	20	8	35	16
Grade Point Average	2.72	3.01	0.89	1.81	3.23	3.74	1.42	2.45	2.9	3.96	1.00	2.53

5. Use the values from the table to construct a scatter plot of the data.

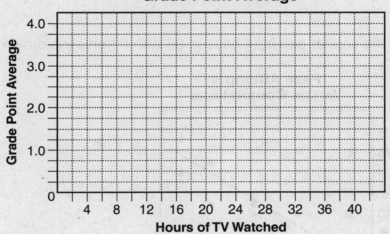

6. Does the scatter plot represent a clustering of data points? _____

7. Is there a relationship between the hours of TV watched by students and their grade point averages? _____

8. Are there any outliers in the data? If so, what are the ordered pairs of the point(s)?

_____

Lesson 36: Trend Lines

The closer the points come to forming a straight line in a scatter plot, the stronger the association between the variables. If you cannot see a line formed by the points, then there is probably no association between the variables. If two variables have a strong association, a **trend line** can be drawn to approximate missing data. A trend line has close to the same number of points above and below it.

Positive and Negative Association

A trend line that increases from left to right represents a **positive association**. A trend line that decreases from left to right represents a **negative association**. If there is no pattern, there is **no association**.

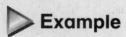 **Example**

Here is the scatter plot from Lesson 35 with a trend line drawn on it.

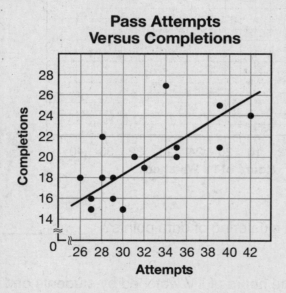

Pass Attempts Versus Completions

The trend line increases from left to right. Therefore, the scatter plot shows a positive association between attempts and completions. The trend line also allows you to make predictions about the data. Looking at the trend line, you can predict that if Brett attempts 37 passes, he will complete about 23 passes.

TIP: A trend line also helps identify the outliers in a data set. The points farthest from the line are outliers, such as (34, 27) in the scatter plot above.

CCSS: 8.SP.1, 8.SP.2

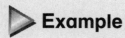 **Example**

A hardware store experiments with different prices for its wagons during different weeks. Depending on the price of the wagons during a week, a different number of wagons are sold. The following table shows the relationship between the cost of the wagons and the number of wagons sold during the week.

Cost of Wagons Versus Number of Wagons Sold

Cost (in $)	60	95	75	80	65	70	100	90	88
Number Sold	30	12	25	20	18	26	10	16	18

The following scatter plot shows the data from the table with a trend line.

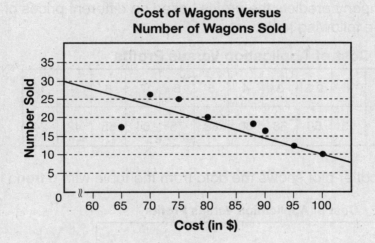

The trend line decreases from left to right. Therefore, the scatter plot shows a negative association between the price of the wagons and the number sold. The trend line also allows you to make predictions about the data. Looking at the trend line, you can predict that if the hardware store charges $85, it will probably sell about 17 or 18 wagons.

The data point at (65, 18) is much lower than the trend line. Therefore, (65, 18) is an outlier in the data set.

CCSS: 8.SP.1, 8.SP.2

Linear and Nonlinear Association

Most scatter plots, where the variables have a strong association, follow a straight trend line. These scatter plots have a **linear association** because the ordered pairs of the data sets follow a straight line. However, some scatter plots show an association with a curved trend line. These scatter plots have a **nonlinear association** because the relationship of the data sets does not follow a straight line. Nonlinear associations may be represented by a curved line.

 Example

A phone application company is deciding how much to charge for its program. The company has decided to charge an amount from $1 to $10. The company knows that fewer people will buy the application if it is more expensive. However, it also knows that some customers won't buy a program if it's too cheap. The company predicts the profits based on different prices of the application in the following table.

Cost of Application Versus Profits

Cost (in $)	1	2.5	3	4	5	6	7	8	9	10
Profits (in $1000s)	15	60	63	80	88	75	64	55	42	24

The following scatter plot shows the data from the table with a trend line.

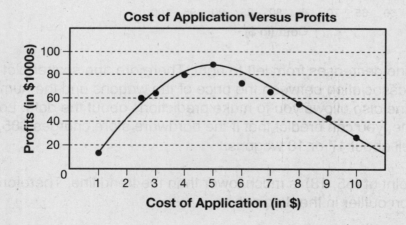

There is an association between the cost of the application and the profits that the company expects to make based on that cost. At first the company makes more money based on the higher cost of the application. But at some point the profits begin to decrease. The trend line forms an upside-down U shape. This is an example of a scatter plot with a nonlinear association.

CCSS: 8.SP.1, 8.SP.2

 Practice

Directions: For questions 1 through 4, write whether each scatter plot shows a positive, negative, or no association. If the scatter plot shows an association, draw a trend line and make a given prediction.

1.

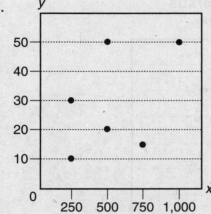

association: _____

prediction for *y* when *x* is 300:

_____

3.

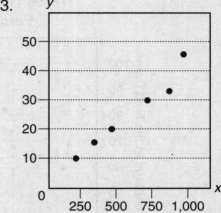

association: _____

prediction for *y* when *x* is 600:

_____

2.

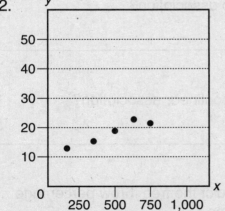

association: _____

prediction for *y* when *x* is 1,000:

_____

4.

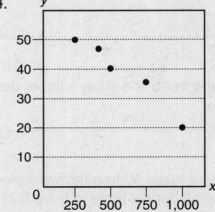

association: _____

prediction for *y* when *x* is 850:

_____

Directions: Use the following information to answer questions 5 through 8.

Joey kept track of the number of free throws that his team shot in practice and the percentage that they made in the next game. He displayed his findings in the scatter plot shown below.

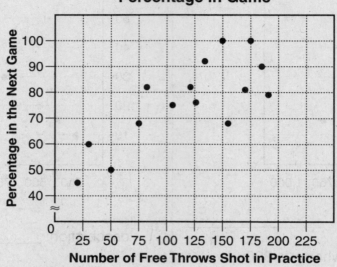

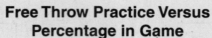

5. Draw a trend line on Joey's scatter plot.

6. Does the trend line show a negative, positive, or no association?

_____

7. Does the trend line show a linear association or a nonlinear association?

_____

8. A student takes 60 free throws during practice. Predict the free throw percentage that the student is likely to have during the next game.

_____

CCSS: 8.SP.1, 8.SP.2

Directions: Use the following information to answer questions 9 and 10.

The following table shows the value of Samantha's car for each of the last seven years that she has owned it.

Car Value Versus Car Age

Age (in years)	0	1	2	3	4	5	6	7
Value (in dollars)	15,800	13,000	11,500	10,800	9,000	8,500	7,500	7,000

Samantha graphed the data from her table in a scatter plot and then drew a trend line on it.

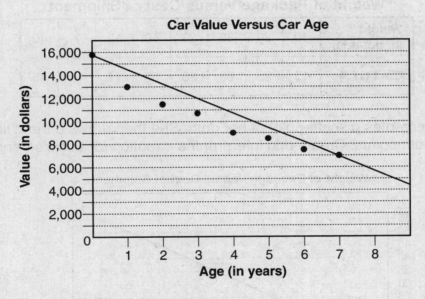

9. Does Samantha's trend line model the data set well?

_____

10. Explain why the trend line does or does not model the data set well.

_____

_____

_____

Lesson 37: Interpreting Linear Models

You can use an equation to describe the trend line on a scatter plot. The equation can be used to find relationships between the two data sets (variables) that are being graphed. The slope of the equation describes the rate of change between the two variables. The *y*-intercept of the equation describes the value of the dependent variable (*y*) when the value of the independent variable (*x*) is 0.

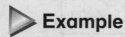 **Example**

Lucia ships eight packages from a delivery company. The weight of her eight packages and the cost to ship them are listed in the following table.

Weight of Package Versus Cost of Shipment

Weight (in pounds)	10	20	30	23	50	12	7	40
Cost (in $)	15	25	35	28	55	17	12	45

Lucia creates a scatter plot of her data. She then draws a trend line through the data points. Create and interpret the equation for Lucia's trend line.

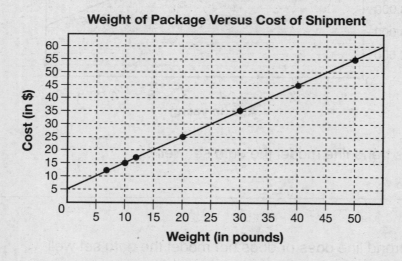

Weight of Package Versus Cost of Shipment

The *y*-intercept of the trend line is 5. The trend line goes up 1 unit for every 1 unit it moves to the right. Therefore, the slope is 1. Lucia's trend line can be modeled by the equation $y = x + 5$, where *y* is the cost to ship a package that weights *x* pounds.

The slope of 1 means that the rate of change is $1 for every additional pound. The *y*-intercept of 5 means that the delivery company charges a minimum of $5 to ship any package, regardless of weight.

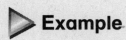

Example

A team of scientists has estimated the population of wolves and caribou in a wildlife preserve during the past 100 years. The wolf is a natural predator of the caribou. The data is given in the following scatter plot, along with the trend line drawn by the scientists.

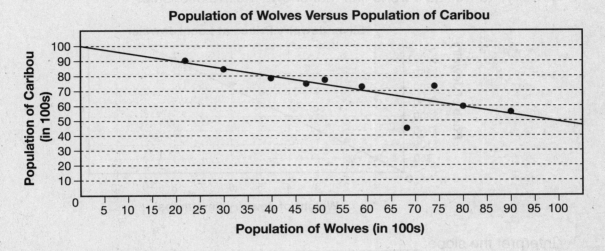

Create and interpret the equation of the trend line.

The team of scientists wants an equation that they can use to predict the population of the caribou or the wolves based on the population of the other species. Create and interpret an equation for the scientists' trend line.

The trend line crosses the y-axis at 10,000. The trend line goes down 1 unit for every 2 units it moves to the right. Therefore, the slope, or rate of change, is $-\frac{1}{2}$. The equation of the scientists' trend line is $y = -\frac{1}{2}x + 10,000$, where y is the population of caribou and x is the population of wolves.

The rate of change of $-\frac{1}{2}$ means that for every 2,000 wolves added to the population, there will be 1,000 fewer caribou. The y-intercept of 10,000 means the caribou population would grow to 10,000 in the nature preserve if there were no wolves.

Practice

Directions: For questions 1 and 2, interpret the slope and *y*-intercept of the trend line for the situation. Then write an equation for the trend line.

1. The following scatter plot shows the prices and weights of several pieces of her jewelry, as well as a trend line that shows their relationship.

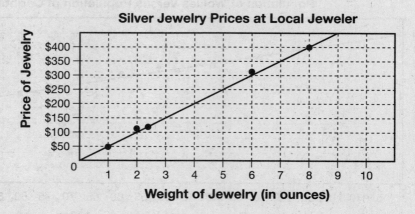

Interpret the slope: _____

Interpret the *y*-intercept: _____

Equation: _____

2. The following scatter plot shows a plumber's charges (not including parts) and the time he spends at each job, as well as a trend line that shows their relationship.

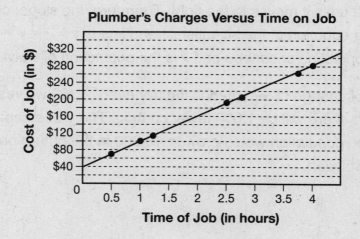

Interpret the slope: _____

Interpret the *y*-intercept: _____

Equation: _____

CCSS: 8.SP.3

Directions: Answer questions 3 through 5 based on the following scenario and scatter plot. A student in Ms. Petricelli's 8th-grade class asked 9 of her friends how far they live from school. She then asked her friends how long it takes them, on average, to get from home to school. The student plotted the data points in the following scatter plot.

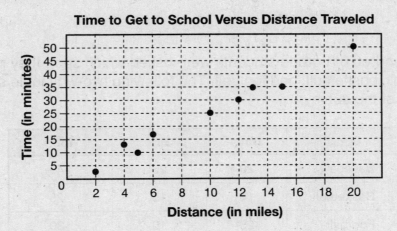

Time to Get to School Versus Distance Traveled

3. Draw a trend line on the student's scatter plot.

   Explain why your trend line is accurate.

   _____

   _____

4. What is the *y*-intercept of your trend line?

   _____

   Explain what the *y*-intercept of the trend line means in terms of the scenario.

   _____

   _____

5. What is the slope of your trend line? _____

   Explain what the slope of the trend line means in terms of the scenario.

   _____

   _____

Lesson 38: Two-Way Tables

A **two-way table** is a table that shows the frequencies of two sets of categorical data. A two-way table can help identify whether there is an association between two variables.

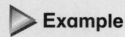 **Example**

One hundred 8th-grade students were asked which sport they prefer among baseball, basketball, and football. The following two-way table shows the results of the survey.

Sport	Boys	Girls	Total
Baseball	20	25	45
Basketball	10	15	25
Football	20	10	30
Total	50	50	100

The two-way table shows the preferred sport for 100 8th graders in total. You can look at the rows to see how the boys and girls voted for each sport. You can look at the columns to see how all the boys voted or all the girls voted—or how many voted for each sport in total.

You can analyze this type of table for specific trends. For example, more students prefer football than basketball in total; however, girls prefer basketball more than football. Similarly, baseball appears to be much more popular than football according to the total in the right-hand column. However, boys prefer baseball equally as much as they prefer football.

You can also create a relative-frequency two-way table based on two sets of categorical data. A relative-frequency two-way table shows the relative frequencies for the whole table, the rows of the table, or the columns of the table. The total relative frequencies are always out of 1, so the relative frequency counts will have values of 1 or less. In some cases they will be rounded to the nearest hundredth.

 TIP: A two-way table often shows the frequency of an event happening. Therefore, two-way tables are also referred to as **frequency tables**. The entries in the table, therefore, may be called **frequency counts**.

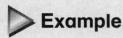 **Example**

Fifty students in the 8th-grade class of a middle school were asked what kind of yogurt they like (vanilla or strawberry) and what kind of topping they like (nuts, chocolate chips, or nothing). Identify any trends in the data.

Topping	Nuts	Chocolate Chips	Nothing	Total
Vanilla	9	8	13	30
Strawberry	7	9	4	20
Total	16	17	17	50

To create a relative-frequency two-way table for all 50 students, divide each number in each cell by 50.

Topping	Nuts	Chocolate Chips	Nothing	Total
Vanilla	0.18	0.16	0.26	0.60
Strawberry	0.14	0.18	0.08	0.40
Total	0.32	0.34	0.34	1.00

To create a relative-frequency two-way table for the rows, divide each number in each row by the total of that row. For instance, for students who like vanilla, divide each number in that row by 30.

Topping	Nuts	Chocolate Chips	Nothing	Total
Vanilla	0.30	0.27	0.43	1.00
Strawberry	0.35	0.45	0.20	1.00
Total	0.32	0.34	0.34	1.00

To create a relative-frequency two-way table for the columns, divide each number in each column by the total of that column. For instance, for students who like nuts, divide each number in that column by 16.

Topping	Nuts	Chocolate Chips	Nothing	Total
Vanilla	0.56	0.47	0.76	0.60
Strawberry	0.44	0.53	0.24	0.40
Total	1.00	1.00	1.00	1.00

◯ Practice

Directions: For questions 1 through 4, answer the question based on the given two-way table.

1. The following two-way table shows the number of students who voted for each of the two candidates for class president, by grade.

Candidate	Grade 7	Grade 8	Total
Zoe	45	20	65
Alessandro	30	60	90
Total	75	80	155

How many more 8th graders voted for Alessandro than voted for Zoe?

A. 15

B. 20

C. 40

D. 80

2. The following two-way table shows the numbers of different colors of cars and SUVs at an auto dealership.

Color	Car	SUV	Total
Red	25	15	40
White	15	10	25
Blue	40	20	60
Green	20	5	25
Total	100	50	150

Which is the least popular color of car in the dealership?

A. white

B. red

C. green

D. blue

CCSS: 8.SP.4

3. The following two-way table shows the number of students who scored an A, B, or C on the first two tests in a class.

Grade	Test 1	Test 2	Total
A	9	6	15
B	15	20	35
C	6	4	10
Total	30	30	60

How many more students got at least a B on Test 2 compared to Test 1?

_____

4. The following two-way table shows the average number of different types of tickets sold in a movie theater for weekday and weekend screenings.

Tickets Sold	Weekday	Weekend	Total
Child	50	100	150
Adult	60	110	170
Senior	30	40	70
Total	140	250	390

How many more children saw a screening at this movie theater during the weekend than adults during the weekday?

_____

Directions: Answer questions 5 through 8 based on the following scenario and two-way table. Round the relative frequency counts to the hundredths, if necessary.

Two hundred students in a summer camp voted for the location of a field trip.

Trip	Zoo	Park	Aquarium	Total
Boys	50	20	30	100
Girls	40	40	20	100
Total	90	60	50	200

5. Create a relative-frequency two-way table for the whole table.

Trip	Zoo	Park	Aquarium	Total
Boys				
Girls				
Total				

6. Create a relative-frequency two-way table for the rows.

Trip	Zoo	Park	Aquarium	Total
Boys				
Girls				
Total				

7. Create a relative-frequency two-way table for the columns.

Trip	Zoo	Park	Aquarium	Total
Boys				
Girls				
Total				

8. Which relative-frequency two-way table would you use to show the percentages of boys versus girls who voted to go to the zoo? Explain why.

_____

_____

_____

Unit 5 Practice Test

Use the following table to answer questions 1 and 2. The two-way table shows the number of students in a middle school who have 0, 1, 2, or 3 siblings.

Siblings	Grade 7	Grade 8	Total
0	10	7	17
1	15	14	29
2	8	18	26
3	7	6	13
Total	40	45	85

1. How many more 7th graders than 8th graders have 1 or no siblings?

   A. 1

   B. 3

   C. 4

   D. 12

2. How many more 8th graders than 7th graders have at least 2 siblings?

   A. 9

   B. 10

   C. 18

   D. 24

3. Ingrid is preparing for a half marathon by running different numbers of miles in the weeks before the race. The following scatter plot shows the number of miles that Ingrid ran and her time in 8 runs before the half marathon.

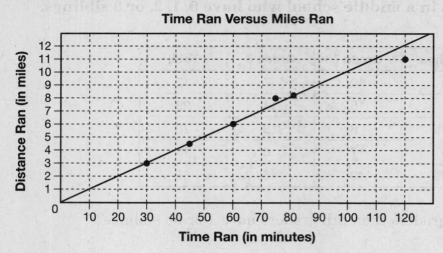

Ingrid does one more run before the half marathon. She runs for a total of 100 minutes. About how many miles will she run during this time?

_____

4. The boiling point of water depends on the altitude where the water is boiled. The following scatter plot shows the boiling points of water at 6 different altitudes, rounded to the nearest degree Celsius.

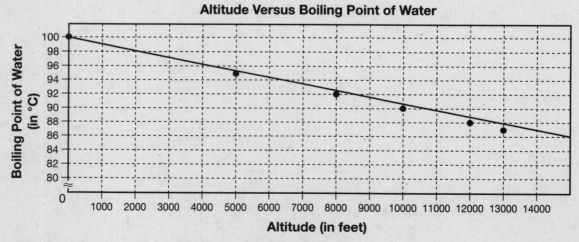

Write an equation that can be used to approximate the relationship between the altitude, x, and the boiling point of water, y.

_____

5. A biologist measures the heights and weights of the 7 adult giraffes at a zoo. The following scatter plot shows the heights and weights of the giraffes.

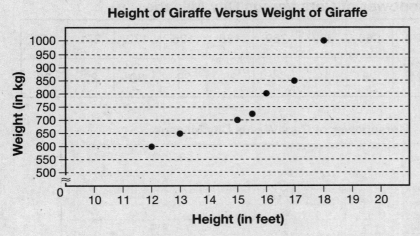

Height of Giraffe Versus Weight of Giraffe

Draw a trend line on the biologist's scatter plot.

6. The following two-way table shows the number of small and large hot dogs that are sold one day at a hot dog stand, as well as the number of hot dogs with ketchup or mustard on them. Each hot dog had exactly one condiment on it.

Condiment	Small	Large	Total
Ketchup	30	30	60
Mustard	10	30	40
Total	40	60	100

Of the large hot dogs sold, 30 had ketchup on them and 30 had mustard on them. Explain whether or not this means that the condiments are equally popular.

_____

_____

_____

7. The city mileage and horsepower of 10 popular cars are given in the following table.

Horsepower of Cars Versus City Mileage

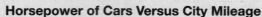

Horsepower	300	270	260	303	269	168	106	170	271	178
City Mileage (MPG)	20	21	22	18	22	25	29	30	22	30

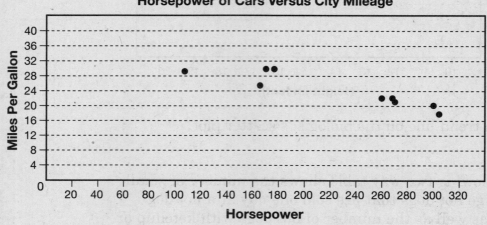

Horsepower of Cars Versus City Mileage

Do the data points in the scatter plot represent a linear or nonlinear association, and do they represent a negative or positive association?

Linear or nonlinear association: _____

Positive or negative association: _____

230

8. Juan recorded the average high temperatures in July for 10 major U.S. cities: El Paso, Jackson, Memphis, St. Louis, New York, Cleveland, Boston, Boise, Seattle, and Juneau. He then determined the degree of latitude for each city. The following table shows the data from Juan's research.

Latitude Versus Average High Temperature in July

Latitude (in °N)	31.86	32.33	35.1	38.64	40.74	41.47	42.35	43.62	47.62	58.3
Avg. High Temperature (in °F)	96	92	92	90	85	82	82	91	75	64

Part A
Create a scatter plot using Juan's data.

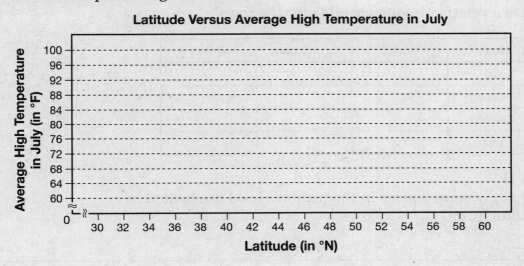

Part B
What are the coordinates of the outlier from Juan's data?

_____

Explain how you were able to identify an outlier from the data set.

_____

_____

_____

9. There are 250 homes in a neighborhood. Some people in the neighborhood live in houses, and some live in apartments. The houses and apartments are heated by oil, gas, or electricity. The following two-way table shows the number of houses and apartments heated by oil, gas, or electricity.

Heat	House	Apartment	Total
Oil	40	10	50
Gas	50	50	100
Electric	20	80	100
Total	110	140	250

Part A
Create a relative-frequency table for the rows.

Heat	House	Apartment	Total
Oil			
Gas			
Electric			
Total			

Part B
How many more houses in the neighborhood are heated by gas than by oil?

_____

Part C
Based on the relative frequencies, which type of fuel is used to heat houses at the same rate that electricity is used to heat apartments?

_____

Explain how you found the fuel that is used to heat houses at the same rate that electricity is used to heat apartments.

_____

_____

_____

Math Tool: Number Lines

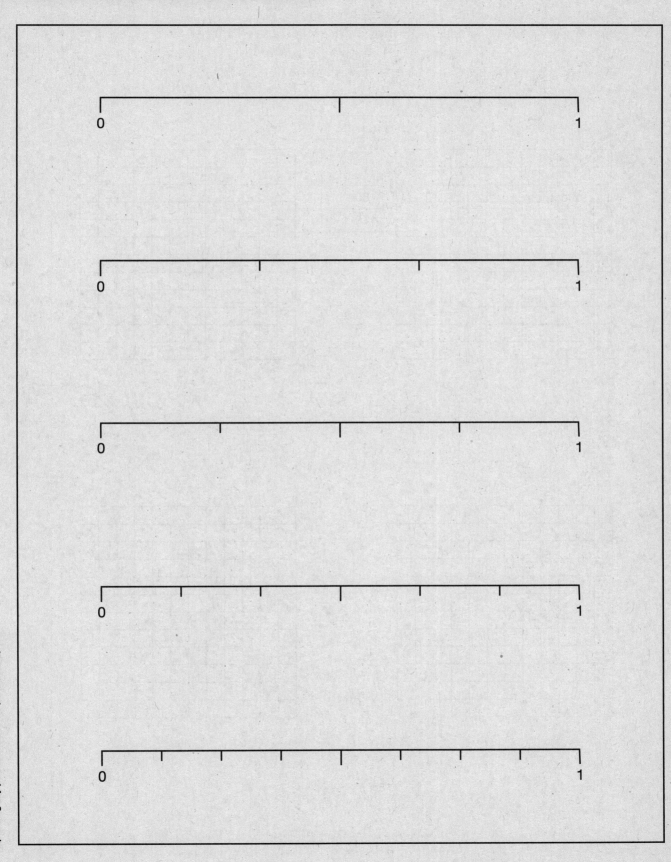

Math Tool: Grids

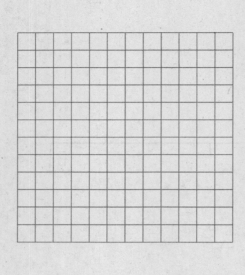

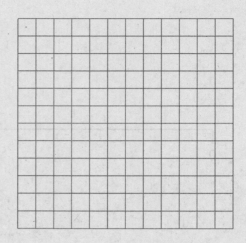

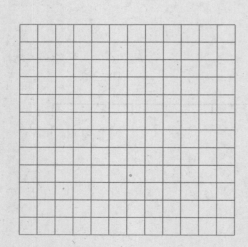

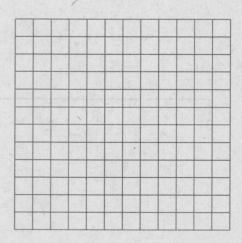

Math Tool: Coordinate Grids

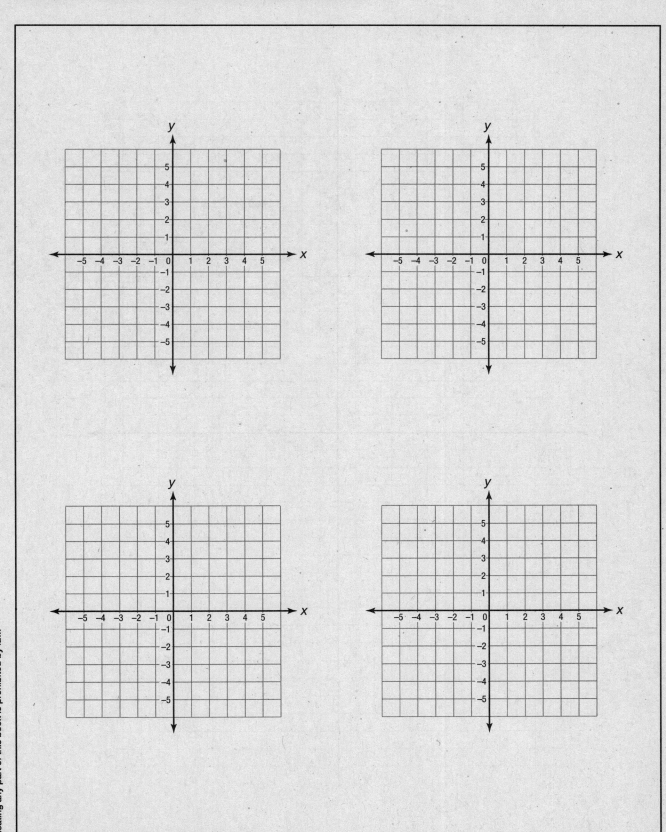

Math Tool: Coordinate Grid

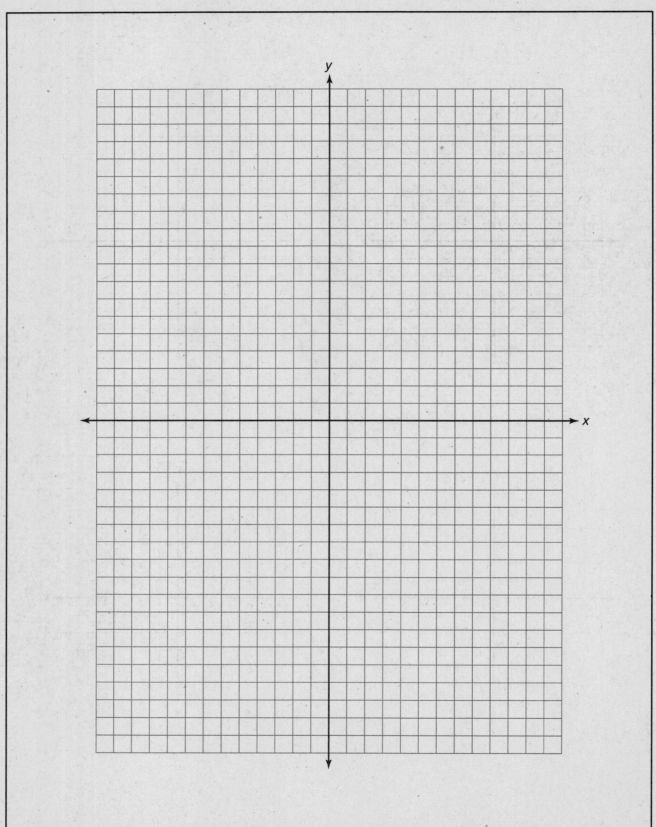

Math Tool: Coordinate Grid

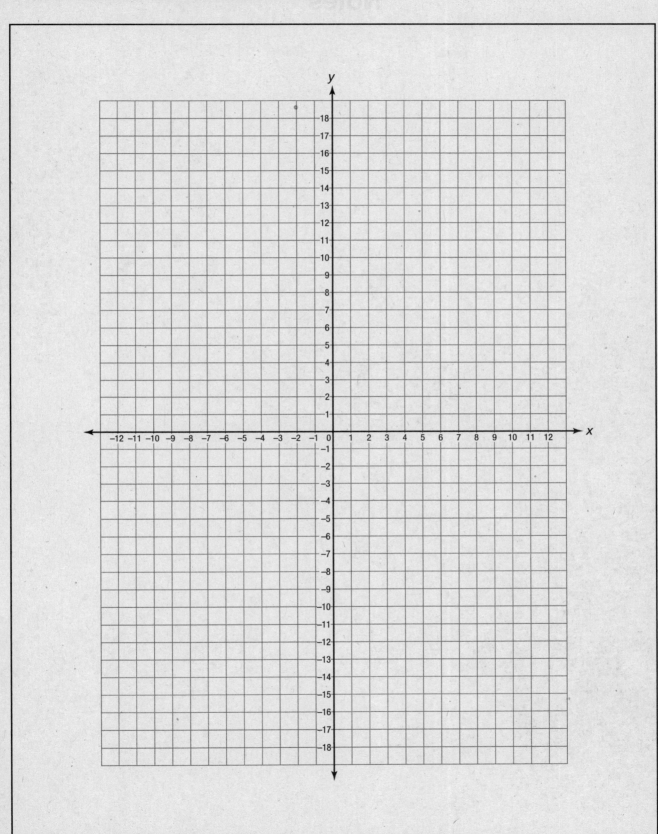

Notes